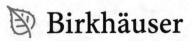

More information about this series at http://www.springer.com/series/4927

Earthquakes and Tsunamis in the Region from Azores Islands to Iberian Peninsula

Edited by
Elisa Buforn · Maurizio Mattesini

Previously published in *Pure and Applied Geophysics* (PAGEOPH),
Volume 177, No. 4, 2020

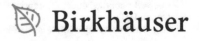 Birkhäuser

Editors
Elisa Buforn
Facultad de Ciencias Físicas
Universidad Complutense
Madrid, Spain

Maurizio Mattesini
Facultad de Ciencias Físicas
Complutense University of Madrid
Madrid, Spain

ISSN 2504-3625
Pageoph Topical Volumes
ISBN 978-3-030-60588-9

This book is published under the imprint Birkhäuser, www.birkhauser-science.com by the registered company Springer Nature Switzerland AG
The registered company address is: Gewerbestrasse 11, 6330 Cham, Switzerland

Contents

Pure Appl. Geophys. 177 (2020), 1695–1697
© 2020 Springer Nature Switzerland AG
https://doi.org/10.1007/s00024-020-02482-4

Earthquakes and Tsunamis in the Region from Azores Islands to Iberian Peninsula

ELISA BUFORN[1] and MAURIZIO MATTESINI[1]

On February 28th, 1969, a large earthquake (M_w = 7.8) strokes the Iberian Peninsula and northern Morocco, producing some casualties and important damage and economic losses. The off-shore focus, located 200 km west from the Cape Saint Vincent (SW of Iberian Peninsula), generated a tsunami that affected the Atlantic coasts SW of the Iberian Peninsula and NW of Morocco. This earthquake is the last large earthquake that has occurred in this region, very near where the catastrophic 1755 Lisbon earthquake took place. Seismicity in this region is associated with the western part of the plate boundary between Eurasia and Africa. From the Azores Islands to Algeria, this plate boundary has a complex nature, consequence of its proximity to the pole of rotation of the African plate. This complexity is shown by changes in the stress pattern from the Azores to Algeria, differences in plate velocities, seismicity rates and occurrence of intermediate and deep depth earthquakes in south Iberia and northern Morocco.

On March 6th–8th, 2019, on the occasion of the 50th anniversary of the 1969 earthquake, a workshop was held in Madrid, organized by the Universidad Complutense de Madrid (Spain) and the Instituto Geográfico Nacional (Madrid, Spain). Two invited lectures and 45 communications were presented covering different aspects of crustal structure, seismic and geodetic instrumentation, seismic archive data, studies of the 1969 earthquake, seismic sources, GPS results, Earthquake Early Warning Systems (EEWS), seismic hazard and tsunami studies. About 100 participants attended, from Europe, Northern Africa, America, and Asia. In this Topical Issue, we have collected ten papers, representative of those presented at the workshop, including the two invited lectures, one paper on crustal structure, three on seismicity and a data archive, three focused on the 1969 earthquake, one on the seismic hazards.

The volume opens with the two invited lectures. The first one, by Allen et al. (2019), presents a description of the MyShake Platform, an operational framework to provide an earthquake early warning. The EEWS is a recent and powerful tool to prevent and mitigate the damage of earthquakes. MyShake uses smartphone technology to detect earthquakes and issue warnings; it has been operating for three years and more than 300,000 people have downloaded it. The paper describes the operational details, distribution, alert generation, and delivery, with a very interesting discussion of results and future directions. The second invited lecture, by Baptista (2019), is a review of tsunamis that occurred along the Azores-Gibraltar plate boundary. The paper starts presenting the tectonic regime along the plate boundary and continues describing the methods used in this study: back-ward ray tracing of tsunami travel times and inversion of tsunami waveforms. The large earthquakes of 1722, 1755, 1761, 1941, 1969 and 1975 are analyzed and a thorough discussion is presented. The author concludes that the six tsunamigenic events studied do not follow along a single plate boundary, confirming previous studies and show an image of the complex lithospheric region.

The paper, by Catalán et al. (2019), is a very interesting study of the plate boundary between Eurasia and Africa at the Gulf of Cádiz and the Canary Archipelago exploring the possible geodynamic connection between them. The study is based on tectonic, thermal, gravimetric and seismic observations. The gravity data and the Curie point depth

[1] Dept. de Geofísica Y Meteorología, Fac. CC. Físicas, Universidad Complutense, 28040 Madrid, Spain. E-mail: ebufornp@ucm.es; mmattesi@ucm.es

show a possible link between the two regions. The authors conclude about the presence of a lithospheric thinning between both regions and of an asthenospheric channel, which feeds and alters locally the plate boundary.

The next three papers deal with the study of the seismicity, historical and instrumental, and about an archive of early seismic documents for the region. The paper by Udías (2019) is a review of large tsunamigenic earthquakes that occurred at SW of Iberia before the 1755 Lisbon earthquake. The author carries out a very detailed analysis of the historical sources with information about the occurrence of earthquakes and tsunamis in the region Cape Saint Vincent-Gulf of Cádiz, separating those occurred before and after the year 500 A.D. He concludes that the 1755 earthquake and tsunami is not an isolated event in SW Iberia and other ones have occurred before the great Lisbon event. The next paper by López Muga et al. 2019 presents the Spanish Geophysical Data National Archive. This center is part of the Instituto Geográfico Nacional (IGN, Madrid, Spain) and its objective is to collect and preserve geophysical information from the IGN observatories. Among the data stored is that corresponding to seismic information (analogical records, seismic bulletins, macroseismic questionnaires, historic photographs, etc....) for the period of 1914 to 1990. As an example of the archive information, the authors show the information available for the 1969 earthquake. The content of the Archive is available for researchers. The last paper of this group, by Cabieces et al. (2020), is a study of the seismicity and focal mechanism of earthquakes that occurred in the SW Cape Saint Vincent-Gulf of Cadiz using an amphibious network (OBS and land stations) during a temporary survey. The hypocenters are located using a nonlinear algorithm and different crustal velocity models, including 3D models, and some cases are analyzed in detail. Focal mechanisms of larger recorded earthquakes have been estimated from first motion polarities and inversion of moment tensor. The authors conclude that the use of the amphibious networks together with a 3D crustal velocity model and non-linear algorithms improve the determination of hypocenters, reducing the depth uncertainties.

The 1969 earthquake was the object of a special interest in the workshop. Buforn et al. (2019), have carried out a re-evaluation of seismic intensities for this earthquake using original documents, some of them not used previously, from Portugal, Spain, and Morocco, obtaining a new intensity map for the whole region affected by the earthquake. Using a new 3D crustal velocity model for the region and non-linear methodology, the 1969 seismic sequence has been relocated, obtaining for most of the foci depths between 30 and 50 km. From the comparison between the observed intensities for 1969 and the large 1755 Lisbon earthquake, the authors propose a different focal mechanism for these earthquakes.

The Aparicio Florido (2019) paper is an interesting contribution to the effects of the 1969 earthquake describing how the earthquake and tsunami were felt on ships at the sea. The observations of these effects on six ships in the Atlantic with different tonnages and located at different epicentral distances and azimuths from the epicenter are given with great detail. It is important to remark that according to this information, the effects of the earthquake and tsunami were perceptible even on deep water. The last paper of this group, by Pro et al. (2020), is a study of the PGV, strong motion and intensity data for the 1969 earthquake using synthetic data to address the lack of strong-motion observations and not-saturated records. To test their methodology, the authors previously generated synthetic ground velocity records for the 2007 and 2009 earthquakes, occurred in the region and with available digital data. By using this approach, they generated synthetic PGV for the 1969 earthquake and from them derived the instrumental intensities for Iberia and northern Morocco. The results show a good agreement between the synthetic and observed intensities.

The last paper, by Fontiela et al. (2019), is an estimation of the human losses and expected damage in Faial (Azores islands), using as scenario the damaging 1980 and 1998 earthquakes, both with magnitude larger than 6.0. The authors used the 2001 and 2011 census to estimate the future population. The damages caused by the 1980 and 1998 shocks are then used to study the distribution of buildings according to the EMS-98 vulnerability classes. Different ground motion prediction equations were tested to choose the more adequate one for the island. Finally, the authors propose two scenarios for future off-shore and inland earthquakes (ranging magnitudes between 6.0 and 6.9), with an estimation of 110–620 dead and 330–1750 injured.

Summarizing, the collected papers in this issue present new results on the structure, seismicity, seismotectonics, seismic hazard, and geodynamics of this complex region from the Azores Islands to the Iberian Peninsula. The Editors would like to dedicate this Topic volume to Dr. Alfonso López-Arroyo (1927–2017) and to Prof. Agustín Udías, for their pioneer contribution on the study of the seismicity, seismotectonics and seismic risk of Iberia. Alfonso and Agustín were two early career geophysicists when in 1972 they published the first study of the 1969 earthquake and its seismic sequence (López-Arroyo and Udías, 1972). Their germinal research efforts laid the foundation for creating an extraordinary research group in Madrid (Spain) that formed and inspired a number of generations of geophysicists. Most of the Authors, which contributed to this Special Issue, were actually former members of this research group or at least indirectly related to it.

Publisher's Note   Springer Nature remains neutral with regard to jurisdictional claims in published maps and institutional affiliations.

References

Allen, R. M., Kong, Q., & Martin-Short, R. (2019). The MyShake Platform: A global vision for earthquake early warning. *Pure and Applied Geophysics.* https://doi.org/10.1007/s00024-019-02337-7. **(the current issue)**.

Aparicio Florido, J. (2019). Effects of the 28th February 1969 Cape Saint Vincent earthquake on ships. *Pure and Applied Geophysics.* https://doi.org/10.1007/s00024-019-02368-0. **(the current issue)**.

Baptista, M. A. (2019). Tsunamis along the Azores-Gibraltar Plate Boundary. *Pure and Applied Geophysics.* https://doi.org/10.1007/s00024-019-02344-8. **(the current issue)**.

Buforn, E., López-Sanchez, C., Lozano, L., Martínez-Solares, J. M., Oliveira, C. S., & Udías, A. (2019). Re-evaluation of seismic intensity for the 28 February 1969 main shock and relocation of larger aftershocks. *Pure and Applied Geophysics.* https://doi.org/10.1007/s00024-019-02336-8. **(the current issue)**.

Cabieces, R., Buforn, E., Cesca, S., & Pazos, A. (2020). Focal parameters of earthquakes offshore Cape St. Vincent using an amphibious network. *Pure and Applied Geophysics.* https://doi.org/10.1007/s00024-020-02475-3. **(the current issue)**

Catalán, M., Martos, Y., & Martín-Dávila, J. (2019). Eurasia-Africa Plate Boundary Affected by a South Atlantic Asthenospheric Channel in the Gulf of Cadiz Region. *Pure and Applied Geophysics.* https://doi.org/10.1007/s00024-019-02380-4. **(the current issue)**.

Fontiela, J., Rosset, P., Wyss, M., Bezzeghoud, M., Borges, J., & Cota Rodrigues, F. (2019). Human Losses and Damage Expected in Future Earthquakes on Faial Island-Azores. *Pure and Applied Geophysics.* https://doi.org/10.1007/s00024-019-02329-7. **(the current issue)**.

López Muga, M., Benayas, I., & Tordesillas, J. M. (2019). Seismic Information at the Spanish Geophysical Data National Archive. Example: The earthquake of February 28, 1969. *Pure and Applied Geophysics.* https://doi.org/10.1007/s00024-019-02340-y. **(the current issue)**.

López-Arroyo, A., & Udías, A. (1972). Aftershock sequence and focal parameters of the February 28th, 1969 earthquake of the Azores-Gibraltar fracture zone. *Bulletin of the Seismological Society of America, 62*(3), 699–720.

Pro, C., Buforn, E., Udías, A., Borges, J., & Oliveira, C. S. (2020). Study of the PGV, Strong Motion and Intensity Distribution of the February 1969 (Ms 8.0) Offshore Cape St. Vincent (Portugal) Earthquake Using Synthetic Ground Velocities. *Pure and Applied Geophysics.* https://doi.org/10.1007/s00024-019-02401-2. **(the current issue)**.

Udías, A. (2019). *Large earthquakes and tsunamis at Saint Vincent Cape before 1755.* A historical review: Pure and Applied Geophysics. https://doi.org/10.1007/s00024-019-02323-z. **(the current issue)**.

(Published online April 20, 2020)

Pure Appl. Geophys. 177 (2020), 1699–1712
© 2019 The Author(s)
https://doi.org/10.1007/s00024-019-02337-7

The MyShake Platform: A Global Vision for Earthquake Early Warning

RICHARD M. ALLEN,[1] ⓘ QINGKAI KONG,[1] and ROBERT MARTIN-SHORT[1]

Abstract—The MyShake Platform is an operational framework to provide earthquake early warning (EEW) to people in earthquake-prone regions. It is unique among approaches to EEW as it is built on existing smartphone technology to both detect earthquakes and issue warnings. It therefore has the potential to provide EEW wherever there are smartphones, and there are now smartphones wherever there are people. The MyShake framework can also integrate other sources of alerts and deliver them to users, as well and delivering its alerts through other channels as needed. The MyShake Platform builds on experience from the first 3 years of MyShake operation. Over 300,000 people around the globe have downloaded the MyShake app and participated in this citizen science project to detect earthquakes and provide seismic waveforms for research. These operations have shown that earthquakes can be detected, located, and the magnitude estimated ∼ 5 to 7 s after the origin time, and alerts can be delivered to smartphones in ∼ 1 to 5 s. A human-centered design process produced key insights to the needs of users that have been incorporated into MyShake2.0 which is being release for Android and iOS devices in June 2019. MyShake2.0 will also deliver EEW alerts, initially in California and hopes to expand service to other regions.

Key words: Earthquake early warning, smartphone seismic networks, earthquake detection, earthquake alerts.

1. Introduction

Seismology is an observational science that has always been limited by our ability to deploy sensing networks to study earthquake processes and the structure of the Earth. Earthquakes continue to have a devastating effect on even the most earthquake-prepared regions of the world, e.g. Japan following the March 11, 2011 M9.1 Tohoku-Oki earthquake. The

Electronic supplementary material The online version of this article (https://doi.org/10.1007/s00024-019-02337-7) contains supplementary material, which is available to authorized users.

[1] Berkeley Seismology Lab and the Department of Earth and Planetary Science, University of California Berkeley, Berkeley, USA. E-mail: rallen@berkeley.edu

MyShake project aims to form a symbiosis between the needs of the seismology research community to collect data for all forms of research, and the needs of society to better mitigate the impacts of earthquakes. MyShake achieves this goal by turning personal/private smartphones into sensors collecting earthquake data, and delivering earthquake information to the user before, during and after an event, including earthquake early warning.

Earthquake early warning (EEW) is the rapid detection of an earthquake, decision about the region to alert, and delivery of an alert to people and automated systems in that region. The development and implementation of EEW has been accelerating with the advance of communications technologies but has been limited to the regions with seismic networks. For reviews of EEW development and implementation see Allen et al. (2009) and Allen and Melgar (2019).

The public warning systems in Mexico (Cuellar et al. 2014; Allen et al. 2017), Japan (Hoshiba 2014), South Korea (Sheen et al. 2017) and Taiwan (Wu et al. 2013, 2016; Hsu et al. 2016) show that the most common and widespread use of EEW is personal alerting for personal protective actions (Allen and Melgar 2019). For many, the response is simply to drop, cover and hold on, but this also includes improving workplace safety for workers in hazardous environments. Other common actions are the automated slowing of trains, opening and closing of pipeline valves and the readying of emergency response equipment and personnel (Strauss and Allen 2016).

Sensing technologies are also becoming much more pervasive, and the concept of the "Internet of Things" describes a world where billions of sensing devices share data in real-time around the globe. For seismologists, the use of accelerometers in a wide range of consumer electronics has driven the development of low-cost devices that the research community has been exploring (Allen 2012). These networks have made use of the MEMS accelerometers in laptop computers, or placed in specially installed boxes that can be easily deployed in homes and offices to detect earthquakes (Cochran et al. 2009; Chung et al. 2011; Clayton et al. 2012, 2015; Minson et al. 2015; Wu et al. 2016, 2019; Brooks et al. 2017).

The MyShake project does not purchase any sensing hardware and does not deploy or maintain any sensors. The sensing hardware is provided by smartphone owners, and the deployment process is facilitated by the Google Play and Apple iTunes stores. This approach removes significant cost while potentially provided access to $\sim$ 3 billion smartphone sensors. This number will likely only grow as smartphones also become the primary means of connecting to the internet in the developing world. The price is that the sensor network is non stationary in every possible way. Users decide to join and leave the network at will by installing and uninstalling the app. The phones move around for some parts of the day, meaning that the vast majority of motions recorded by the accelerometer are not earthquakes. So, the system must effectively filter this out. MyShake must also achieve this while not impacting the normal daily use of the phone. Most notably this means that the app must use minimal power.

There are other earthquake related projects that use smartphone apps or related technologies such as social media and other forms of internet data collection. Perhaps the most familiar is the USGS "Did You Feel it?" (DYFI) which collects user felt reports via the internet (Atkinson and Wald 2007; Wald et al. 2012). The use of twitter messages sent by users experiencing earthquakes has also been explored (Earle 2010). At the European Mediterranean Seismological Centre (EMSC) multiple sources of crowdsourced information are combined to detect earthquakes. The EMSC LastQuake app provides earthquake information and collects similar experience data to DYFI. By monitoring for increased traffic to their website, use of their app and twitter, the EMSC can rapidly recognize and locate earthquakes (Bossu et al. 2018; Steed et al. 2019). The

Earthquake Network app uses the smartphone accelerometer to detect sudden movements of a phone and sends these trigger times and locations to a server which then attempts to detect and locate earthquakes based on clusters of triggers (Finazzi 2016). The Earthquake Network and MyShake apps are the only apps that use smartphone accelerometers to detect earthquake ground shaking. What is unique about MyShake is that the on-phone app attempts to distinguish earthquake ground motions from everyday motions and records 5 min of accelerometer data when the motion is classified as an earthquake.

In this paper we present an overview of what we refer to as the MyShake Platform. This is a technological framework to facilitate EEW in multiple regions around the world. The MyShake Platform can provide a smartphone-based sensing network to generate alerts, or can receive alerts from other sources, e.g. regional seismic networks. The Platform can distribute these alerts to smartphones or push them to other dedicated alerting systems. We first summarize the current status of the global MyShake network, and then describe the functional components of the MyShake Platform. We then review the operational lessons from the first three years of MyShake and describe how these lessons led to the development of the new MyShake system that we refer to as MyShake2.0.

2. The MyShake Global Smartphone Seismic Network

The MyShake smartphone app turns personal/private smartphones into seismic sensors. Smartphone owners must first download the free app from the Google Play store (and now from the Apple iTunes store). Once installed on a phone, the app registers with the MyShake servers that operate in the cloud. The phone then becomes a sensor that is part of the MyShake global seismic network.

Continually monitoring the accelerometer or streaming continuous data requires too much power and bandwidth to be practical. Instead, an individual phone either needs to trigger or be triggered to record for short periods of time. The key technology that makes the MyShake seismic network possible is an artificial neural network (ANN) embedded in the smartphone app that distinguishes between earthquake-like ground motions and everyday motions (Kong et al. 2016b). In testing, the algorithm was able to correctly identify earthquake-like from other types of ground motion more than 90% of the time. To detect an earthquake, first a phone must be stationary. It then starts to monitor the accelerometer for sudden motion using a STA/LTA trigger (Allen 1978). When the phone then moves, the ANN algorithm classifies the motions as earthquake-like or not. If earthquake-like, a trigger message including the time, location and waveform characteristics is sent to the MyShake servers for use in real-time. The phone also records a total of 5 min of 3-component acceleration data which is later uploaded to the server for archival and later analysis. By using a 1-min ring buffer, the waveforms start 1 min before the trigger and continue until 4 min after.

The first public version of the MyShake smartphone application was released in February 2016. Public interest in the app was significantly greater than expected and over 335,000 people around the globe have installed the app to date (Fig. 1).

MyShake phones have detected over 900 earthquakes with magnitudes from M1.6 up to M7.8 (Kong et al. 2016a). Figure 2 shows an example of the distribution of MyShake phones triggered by a M4.4 earthquake. The 5-min waveforms can be used for a variety of purposes. These include traditional regional seismic network type of operations such as earthquake detection, location and magnitude estimation (Kong et al. 2019). The dense arrays of phone sensors can also be used to generate ShakeMaps. The recorded MyShake data captures both the response of the earth to earthquake excitation and also the response of the buildings. This means that in addition to earthquake source parameters, the characteristics of the buildings can also be determined (Kong et al. 2018). Finally, phones can be remotely triggered to record for short periods of time, which allows the phones to be used as an array. Initial evaluation of the capability suggests that the MyShake network could detect and locate earthquakes as small as M1 using this approach (Inbal et al. 2019). This is smaller than the events that traditional regional seismic networks can typically detect and could assist in the

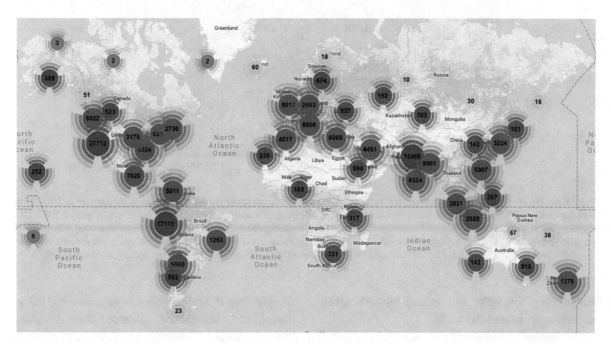

Figure 1
Map illustrating the global distribution of MyShake usage. Plotted are the locations of MyShake phones at the time they register. The locations are gathered into clusters (colored dots) and the number of phones in the cluster is shown

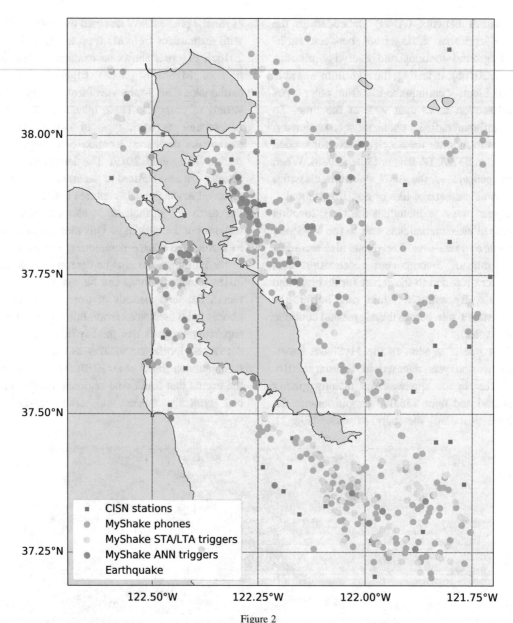

Figure 2
Map of the San Francisco Bay Area showing the locations of the 593 active MyShake phones at the time of the January 4, 2018, M4.4 Berkeley earthquake (colored dots). At the time of the earthquake, 264 of the stationary phones triggered due to sudden motion (STA/LTA trigger—yellow dots), and the ANN algorithm recognized the shaking as earthquake-like on 52 of these phones (orange dots). The ability of the ANN algorithm to recognize earthquake-like shaking reduces rapidly for earthquakes below M5.0 which is the reason the ratio is low for this earthquake. Also shown are the 54 traditional seismic sensors of the CISN that contribute to the ShakeAlert warning system

identification of faults and hazard beneath urban environments around the world.

While the MyShake network is a global seismic network that records data for a range of purposes, earthquake early warning has always been a primary motivation and goal for the network. The reason is simply that if MyShake wants to make use of an individuals' smartphone resource, then MyShake needs to fulfill a need for that user. While many users downloaded earlier versions of the app simply to participate as citizen scientists and obtain basic earthquake information, providing a warning allowing a user to brace and protect themselves in the seconds before earthquake shaking is a good reason for anyone in an earthquake prone region to download the app.

3. *The MyShake Platform for Earthquake Early Warning*

Any earthquake early warning system requires a detection network, a data analysis and alerting decision module, and finally an alerting network (Fig. 3). Since the inception of EEW systems, alerts have been delivered to cellular devices, meaning that smartphones are a natural mechanism to deliver alerts. MyShake also creates the capability of detecting earthquakes and generating an alert using the sensors in smartphones. The "MyShake Platform" provides an operational framework to deliver EEW alerts to smartphones that can be generated by either phone-based detection or using traditional regional seismic networks. Here we outline the components of the MyShake platform and illustrate how MyShake can operate as an end-to-end early warning system or

could interface with traditional earthquake detection system to deliver and possibly enhance alerts.

The smartphone-based earthquake detection starts with individual MyShake phones triggering on an earthquake and sending the trigger information (time, location and peak amplitude) to the MyShake server. The server then looks for space–time clusters of triggers to confirm that an earthquake is underway. This is achieved by dividing the Earth up into 100 km^2 grid cells. For a cell to be "activated" multiple phones in that cell must simultaneously trigger. The criterial for a cell to activate is that there must be a minimum number of stationary phones currently monitoring, and then a defined fraction of them must individually trigger.

Once more than two cells have activated, a modified version of the "density-based spatial clustering of applications with noise" algorithm (DBSCAN, Ester et al. 1996) is used to recognize clusters of activated cells. When a cluster is found by DBSCAN, an earthquake is declared. DBSCAN can create any number of earthquakes around the globe simultaneously and associate newly triggered phones to an existing earthquake as time progresses. The earthquake is located based on the individual phone

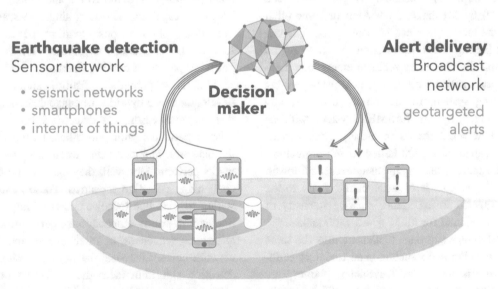

Figure 3
Schematic figure showing the components of the MyShake Platform. All earthquake early warning systems must have a sensor network to detect earthquakes (left), a decision maker that analyses the data and decides on when to alert a specific region (center), and an alert delivery network (right). MyShake can use smartphones for all of these tasks. It can also interface with other sources of alerts or deliver alerts through other non-smartphone distribution networks

trigger times using standard techniques. The magnitude is estimated based on the peak amplitude observed by the phones at the time of the trigger. We use a random forest regressor to estimate the magnitude. This approach to event classification is described in detail in Kong, Martin-Short and Allen (in review), and is currently undergoing testing on the real-time system. What has become clear over the first few years of MyShake operation is that the MyShake network is very heterogeneous and dynamic i.e. time-dependent. As the network grows, the earthquake detection algorithm will undoubtedly have to develop with it.

The MyShake platform can receive earthquake source information from any source, not just from the MyShake phone network. For example, the MyShake platform currently receives an earthquake feed from the ShakeAlert system in the United States. ShakeAlert uses traditional seismic network stations to detect earthquakes, locate them and estimate their magnitude (Chung et al. 2019). It then generates "ShakeAlerts" for events greater than M3.5 in its operational region of California, Oregon and Washington (Kohler et al. 2017).

Once an earthquake has been detected and characterized (typically location, origin time and magnitude) by MyShake, ShakeAlert or some other system, the alert region can be determined. The criteria for issuing alerts varies from region to region and is typically chosen by regional emergency managers in consultation with the seismologists operating the warning system. In Mexico for example, any event detected by the SASMEX system with an estimated $M > 6.0$ results in alerts across various cities. In Japan, alerts are issued to sub-prefectures when the expected shaking intensity exceeds 5-lower. In the US, the goal is to issue public alerts to the region expected to experience shaking intensity of MMI 3 and greater for all $M \geq 4.5$ earthquakes.

The MyShake approach to alerting uses the same 10 km by 10 km cell structure used to detect earthquakes to issue alerts. MyShake phones are dynamically registered to one of these 100 km^2 cells. This provides sufficient location accuracy for geo-targeted alerts, while preserving the privacy of users. When an earthquake detection is reported to the system, the alert region is defined based on the desired criteria for the region. In the case of the US, a contour is drawn around the region expected to experience MMI 3 or greater. All phones registered to a 100 km^2 cell that is within or overlaps with the contour are then alerted.

The MyShake platform provides delivery of alerts to phones using standard push notification protocols provided by Google and Apple for delivery to Android and iOS apps, respectively. The alert messaging pathway starts from the MyShake servers to Google/Apple, and then on to cellular providers or through WiFi depending on how the phone is connected to the internet. The pathways to the phones are highly varied. In some cases, the MyShake servers send messages directly to individual phones. In other cases, a single message is sent to phones grouped by their coarse location and that message is directed to individual phones by a later step in the pathway. In all cases, the last step of the alert messaging pathway involves sending a message to each phone over a network connection. This raises the question of how regional cellular and WiFi systems will handle a message intended for a large number of phones.

There is no reason why an alert from the MyShake Platform could not also be delivered using other protocols. In South Korea and Japan, a cellular broadcast capability allows a single message to be "broadcast" simultaneously to all phones in a region. In Mexico City, sirens distributed across the city notifying people of coming shaking.

Finally, a potentially fruitful area for future development is a hybrid earthquake detection method that integrates earthquake detections from traditional seismic networks with phone-based triggers. While the detection methods using traditional seismic networks perform very well, they can still be fooled in challenging detection scenarios. These include false station triggers due to noise spikes or large teleseismic events, poorly associated triggers during intense aftershock sequences, or other unexpected forms of network noise. Likewise, the MyShake phone-based detection algorithm faces the challenges of heterogeneous phone distribution and the need for relatively high signal levels to enable detection.

Figure 2 shows the traditional seismic stations of the California Integrated Seismic Network (CISN) that are used for ShakeAlert, along with the MyShake phones operating at the time of the January 4, 2018, M4.4 Berkeley earthquake. ShakeAlert issued a warning when the 5 CISN stations closest to the event triggered. By the time the ShakeAlert server received these 5 triggers, the MyShake server had received 28 phone triggers. The 28 phone triggers were distributed across a much wider area than the 5 traditional sensors. Both the larger number and wider footprint of detections makes an alert that uses both networks more robust, and potentially faster.

4. Initial Operation of MyShake

In this section we review some of the operational details, challenges and observations from the first 3 years of operation. We cover how the app is distributed and what drives downloads. We then provide examples of earthquake detection by MyShake smartphones and illustrate the capabilities to detect and generate alerts. Finally, we provide some data on the speed with which MyShake can deliver alerts to users.

4.1. Network Distribution

The primary advantage of MyShake over other seismic networks is the ease with which we can "deploy" seismic sensors. A smartphone owner simply has to decide to download the app from an app store and install. The challenge is to make smartphone owners aware of the MyShake app and want to download it.

The history of MyShake app usage from initial deployment in February 2016 through June 2019 is shown in Fig. 4. Note that this history precedes the release of MyShake2.0 (discussed later), and so from a user perspective the app did not change. This initial version of MyShake that we report on was only available for Android and provided limited features for the user. These included a map of recent earthquakes, basic safety information and some information about past earthquakes of historical significance. It provided no alerts, nor any information about the earthquakes detected.

While users are downloading the MyShake app on a daily basis, interest and downloads was primarily driven by two types of events: earthquakes and media coverage. The initial release of the app was coincident with the publication of the first MyShake paper describing the earthquake recognition algorithm and was announced with a press conference. There was strong media interest and over 300 news stories were published about the app in the first few days. After 2 days the number of registered phones exceeded 50,000 and within 2 weeks it exceeded 100,000.

The number of "active" MyShake phones is the number that has provided data to the servers within the last 24 h. It peaked 2 days after the launch at 25,000. This immediately illustrates the central challenge for MyShake: how to keep users interested so that they keep the app installed and operational on their phone for long periods of time.

Following the launch, there have been four events resulting in step-function increases in the number of users (Fig. 4). The first was May 22, 2016 when a Japanese language version of the app was released. This elicited almost no interest in Japan, but generated media interest in India where the majority of the 11,000 new downloads occurred. The second was in December 2016 and was due to a New Scientist article about the app which generated 5000 downloads, followed by media interest at the American Geoscience Union meeting that resulted in 20,000 downloads. This also took the total number of registered phones past 200,000.

The third step increase was driven by the earthquakes in Mexico in September 2017. The sequence started with a M8.1, then an M7.1 that did most damage, and finally M6.4 that generated a SASMEX warning (Allen et al. 2018). In all this resulted in about 6000 downloads. Finally, the fourth was the M8.0 beneath Peru on May 26, 2019. While it caused little damage due to its 135 km depth, it generated 15,000 downloads taking the total number of registered phones past the 300,000 mark.

4.2. Earthquake Detection

Empirical observations of the deployed MyShake network show that the on-phone earthquake detection algorithm is able to trigger and recognize earthquake ground motions from M5 earthquakes out to ~ 250 km. A M4 can be detected out to ~ 150 km, and a

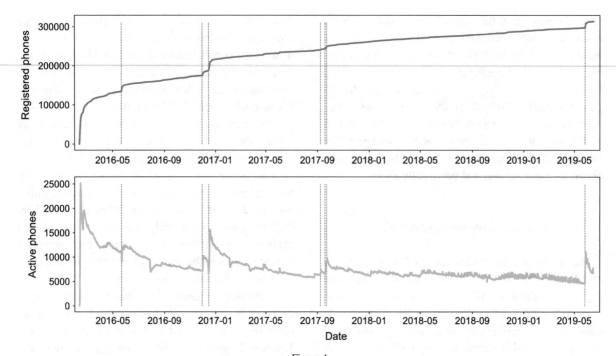

Figure 4
Plot showing the total number of registered phones (top) and the number active phones (bottom) since the launch of MyShake. The number of active phones is the number that have connected to the MyShake server within the last 24 h. The spikes in the number of active phones (vertical dashed lines) follow significant earthquakes and/or media coverage about MyShake

M3 out to $\sim$ 50 km (Kong et al. 2019). With this detection capability, MyShake recorded ground motions from over 900 earthquakes in the first 2 years of operation.

The capability of MyShake to characterize the hypocentral parameters of the earthquake (location, magnitude and origin time) is dependent on the number and geographic distribution of MyShake phones around the event, just as it is with traditional networks. In the first 2 years of operation there were 21 earthquakes for which there were 4 or more seismic phases (P- or S-waves) detected that also had an azimuthal station/phone gap less that 180°. Using Hypoinverse, Kong et al. (2019) showed that the epicenter locations determined from the MyShake data had a median difference to that reported in the USGS Comcat of 2.7 km with a standard deviation of 2.8 km. The depth was different by 0.1 $\pm$ 4.9 km and the origin time was different by 0.2 $\pm$ 1.2 s. Using an M_L relation based on Bakun and Joyner (1984) the median magnitude difference was 0.0 $\pm$ 0.2.

Figure 2 shows the detection of the January 4, 2018 M4.4 Berkeley earthquake. At the time of the event (2:40 am local time) there were 593 active MyShake phones in the Bay Area. Of these, 264 of the stationary phones triggered due to the sudden motion of the earthquake. The ANN algorithm then recognized the ground shaking as being due to an earthquake on 52 of the phones. The ANN algorithm was trained on M5.0 and larger earthquakes, so its ability to recognize shaking in smaller earthquakes is limited. While this earthquake is well recorded, it is at the lower end of the magnitude range for which we expect MyShake to perform well. The best recorded earthquake to date was the June 10, 2016, M5.2 Borrego Springs earthquake in Southern California that triggered over 200 MyShake phones and produced over 100 good seismic waveforms that show clear P-wave arrivals out to $\sim$ 100 km (Kong et al. 2016a).

4.3. Alert Generation

MyShake has not issued any alerts to date. However, the accumulated database of real-time earthquake detections and acceleration waveforms has been used to develop a new algorithm to rapidly characterize earthquakes. It is specifically tailored to the information provided by the MyShake phone detections so that an alert could be issued and limited to the geographic region expected to feel shaking. Using the MyShake observational archive, a simulation platform has been created to mimic the alert performance of a MyShake network. Thousands of simulations were then run with many instances for each earthquake using catalogs of past earthquakes around the world. The alerting algorithm and simulation platform are described in detail in Kong, Martin-Short and Allen (in review).

Here we illustrate the possible performance of MyShake-generated alerts using the September 28, 2018, M7.4 Palu earthquake on the Indonesian island of Sulawesi using the simulation platform. This earthquake and tsunami killed over 2000 people in a region prone to earthquakes. It is estimates that more than 40% of the population owns a cell phone in Indonesia. This example was generated using the simulation platform assuming that 0.1% of the population have MyShake on their phones. We use this number as this is the fraction of the population that has downloaded MyShake in the Los Angeles region. It is therefore a larger number than the current number of active phones, but also seems like a realistic goal for an operational early warning system.

Figure 5 shows snapshots from the simulation, a second-by-second movie of the simulation is included in the supplemental materials. The simulation shows that individual MyShake phones start to detect the earthquake 3.8 s after the origin time, which is when the P-wave first arrives at the surface. At 5.2 s after the origin time, when 26 triggers have been reported, the system recognizes the event, locates it, and estimated the magnitude to be M7.2. This is the point in time when the first alert could be issued across the region. It is 15.6 s before the S-wave reaches the city of Palu where the population is densest and is before the S-wave has reached the surface of the Earth. Based on the estimated magnitude and location, the predicted shaking intensity in Palu is MMI 7. The observed intensity was MMI 8.

4.4. Alert Delivery

While MyShake has not yet delivered any alerts that are visible or apparent to the users, MyShake has sent silent alerts to MyShake phones to evaluate the delivery latency. Figure 6 shows the distribution of alert latency for a test in which a silent alert was sent to just over 100 MyShake phones in the San Francisco Bay Area. The alert latency is the time from when the alert is sent by the MyShake servers to the time that it is received by the phone. This information is reported by each phone receiving the alert. The media delay was 2.8 s. This test was done using the current Google Firebase messaging protocol.

It is unclear how this latency will scale with the number of phones, but it should be expected to increase. The alert delivery pathway starts with the MyShake servers identifying which phones (based on location) should receive the alerts. The message is then passed to Google or Apple, who then process and pass the message to the smartphone service provider who deliver it to a specific phone. The total latency is therefore dependent on the infrastructure of multiple groups. While it would be ideal for alerts to be delivered in less than a second, we should be accepting of larger delays initially. A 2018 test of the US Wireless Emergency Alerts (WEA) system operated by the Federal Emergency Management Agency (FEMA)—commonly known as Presidential or Amber alerts—had a median delay of 13 s as determined by the ShakeAlert project (Douglas Given, personal communication). Despite this delay, the California Office of Emergency Management (CalOES) has determined that this is still a useful path to deliver ShakeAlerts. Thus, while we would prefer the MyShake alert delivery to be faster than 2.8 s, it is still likely useful if it takes a little longer.

5. MyShake2.0: Improved Features and User Interface

MyShake is built on the participation of users. Its success as a global network and its utility for users themselves is reliant on user engagement and continued involvement. In recognition of this fact, we have taken a human-centered design approach to

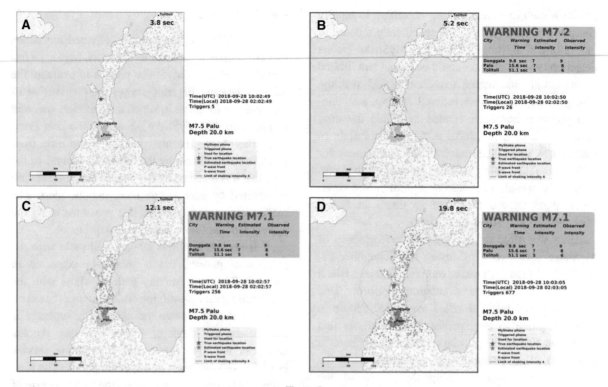

Figure 5

Snapshots from the simulation of the September 28, 2018, M7.4 Palu earthquake in Indonesia. The green dots indicate simulated MyShake phones and the blue star is the epicenter. Phone locations represent a randomly selected sample equivalent to 0.01% of the population. The population is densest in the city of Palu. **a** 3.8 s after the origin time. The P-wave reaches the surface (yellow circle) and starts to trigger some of the MyShake phones (orange dots). Note the two orange dots some distance from the earthquake. These are random phone triggers that are also modeled by the simulation platform. **b** 5.2 s after the origin time. A total of 26 phone triggers (orange dots) have reported at this time, and the network detection algorithms recognizes an earthquake is underway. It is located (light blue star) and the magnitude estimated to be M7.2. This is the first point in time when an alert could be issued of coming ground shaking across the region. The estimated shaking intensity in Palu is MMI 7 (actual intensity was MMI 8) and it is 15.6 s until the S-wave will arrive in Palu. The S-wave has not yet reached the surface of the Earth. **c** 12.1 s after the origin time. This is when the P-wave reaches the center of Palu. The inner yellow circle is the S-wave. Many additional phones have triggered, and the location and the magnitude updated. **d** 19.8 s after the origin time. Approximately when the S-wave reaches the center of Palu. The full simulation results are available in the movie file included as an electronic supplement (Allen-MyShake-Palu_0.001.mp4). Additional examples of actual and simulated performance are included in Kong, Martin-Short and Allen (in review)

proposing and developing the features to be included in the next generation of MyShake. We have also redesigned the user interface. We engaged with MyShake users to understand user needs, behaviors, and pain points in order to identify what features could be added to the app to address these needs. A description of the entire design process can be found in Rochford et al. (2018). Here we highlight some of the key learnings and how they have been mapped into the new MyShake2.0.

We identified four key insights that broadly represent how the public thinks and reacts to earthquakes based on interviews (Rochford et al. 2018):

1. Participants expressed feelings of fear toward earthquakes, as well as feeling helpless to prepare themselves for a large event.
2. Participants reported avoiding thoughts of earthquakes, only engaging with earthquake and preparedness information when they either experienced one or saw one in the news.
3. Respondents wanted information about how an earthquake affected the areas where their loved ones and homes were.
4. Respondents relied on several different resources for news after an earthquake, including social

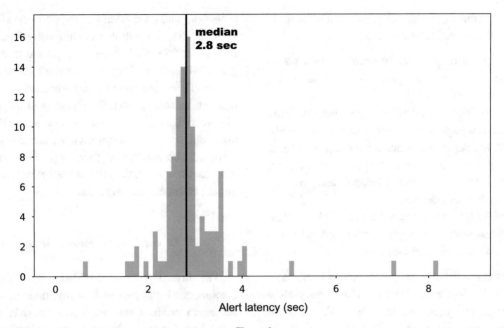

Figure 6
Observed alert latency. This is the time from when the MyShake servers sent the alert until it was received by the phones

media, local news, search engines, and the United States Geological Survey website.

These interviews also lead to the development of a flow of actions that people took, or thought they would take, when an earthquake occurs (Fig. 7). This starts with recognition that an earthquake is underway, moves into a confirmation phase using information from various sources. Then an assessment phase were people are seeking information about what just happened and the impact. Finally, a secondary information seeking phase when people are looking for official advisory information about what to do next.

Based on the analysis of people's needs in an earthquake, and also on their reaction to participating in the MyShake project as citizen scientists, we

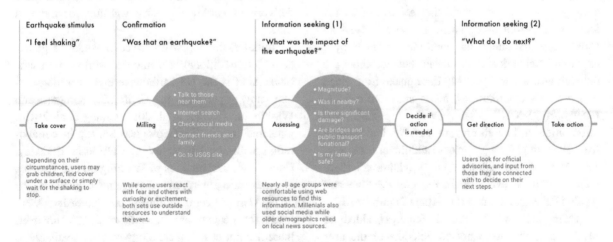

Figure 7
Flow of actions taken by people in response to an earthquake derived from interviews with MyShake users and experts (Rochford et al. 2018)

identified features that were of greatest value to users and were also technically feasible:

1. Delivering earthquake early warning when possible so users can take protective actions in the seconds before shaking.
2. Providing highly local information about building and road damage so users can assess what actions, if any, to take in the minutes after an earthquake.
3. Proving resources that empower users to learn about earthquakes and preparedness, without inciting fear and despair.
4. Providing information about a user's contribution to citizen science through metrics and acknowledgement of their participation.

All of these features have been added to the new version of MyShake while also completely redesigning the user interface to make it both attractive and compliant with best practices in usability. We refer to this new version of the apps (Android and iOS) and system as "MyShake2.0".

MyShake2.0 will issue and receive earthquake early warning alerts. These are geotargeted alerts sent to specific phones that are located within a defined warning area. The app provides a simple interface for users to report their experience and observations when they are in an earthquake. These "experience reports" include the shaking intensity, observed building, road and bridge damage. This information is aggregated with reports from all users and is immediately displayed in the app so users can see where damage has occurred around themselves and in the locations of loved ones. There is also a "MyShake Data" page that shows the contributions made by a user to the network and also basic statistics about the network as a whole. The "My Earthquake Log" page lists a user's contributions, alerts and notifications. Finally, a "Safety" page provides straightforward information on how to prepare, survive, and recover from an earthquake with links to additional resources. The Android and iPhone apps providing all these features will be available in the Google Plan and Apple iTunes store when this article is published.

There are two additional features added to MyShake2.0 that were not an outcome of the user engagement process described above. Both are added to allow for seismological education activities in the classroom and have been developed in collaboration with the IRIS Education and Outreach program. The first is a "Sensor" page that provides a simple graphical readout of the 3-component accelerometer. It also allows this output to be recorded to a csv file that can be shared from the phone to allow analysis. The second is a simplified version of IRIS's StationMonitor [https://www.iris.edu/hq/inclass/software-web-app/station_monitor]. This allows a user to browse earthquake waveforms from traditional/permanent seismic stations.

6. Discussion and Future Directions

The first 3 years of MyShake operations have demonstrated that personal smartphones can be harnessed to produce seismic waveform data that has value both for research and hazard reduction. While the response of the citizen science public to the first version of the app has been stronger than expected, it is clear that the MyShake app must be installed on many more phones in many more regions if it is to generate the necessary data to provide early warning. For this to be possible MyShake2.0 must provide greater value to users than the first version of the app.

At the time of writing, MyShake2.0 is in the final stages of testing with an expected release in June 2019. As described above, it provides a much greater range of features that have been designed specifically to address the needs of users. These will include delivery of earthquake early warning for the first time.

Initially, the app will deliver ShakeAlerts generated by traditional seismic networks to users. ShakeAlert is the US earthquake early warning system now operating in California, Oregon and Washington (Kohler et al. 2017; Chung et al. 2019). Operated by the US Geological Survey in collaboration with the University of California Berkeley, Caltech, the Universities of Washington and Oregon, and the state emergency management agencies, the ShakeAlert project determined that ShakeAlerts were ready for general use in October 2018. However, there has not been a technology solution available to deliver alerts to the broad public across all three states. The City of Los Angeles developed the

ShakeAlertLA app that delivers alerts to the public, but only for earthquakes in the County of Los Angeles.

MyShake2.0 will initially deliver ShakeAlerts to an expanding number of test users. But, sponsored by the California Office of Emergency Services, it will deliver ShakeAlerts to the public before the end of 2019. Through this partnership with ShakeAlert, MyShake can simultaneously start delivery of alerts across the region filling ShakeAlert's technology gap, while also expanding the number of users and testing the capabilities of MyShake to generate alerts. There is no plan to issue MyShake-phone-generated alerts in the US, however we will explore how MyShake-phone triggers could be used to enhance ShakeAlert for faster and more reliable alerts.

The next step for MyShake will be to start delivering phone-generated alerts in a region or regions where the density of MyShake phones is sufficient to provide reliable alerts. This will require partnership with the appropriate public safety and emergency management agencies in the region in order to ensure appropriate messaging and educational strategies to inform users.

While there remain substantial technological challenges to global delivery of earthquake early warning, our hope is that MyShake can drive the development of the necessary technologies. As MyShake capabilities improve and become more widely known, components of the MyShake Platform could be integrated with other technologies. Other popular apps could embed the MyShake detection capability within their app thus providing many more sensors without users needing to know specifically about the MyShake app and installing it. Likewise, the alerting component could be integrated into existing popular apps. An extension of this approach would be for the MyShake capability to be directly integrated into smartphone operating systems. As such, the ultimate success of the MyShake effort may be that the MyShake app is no longer needed.

Acknowledgements

The MyShake team at UC Berkeley who all contribute to this work includes Stephen Allen, Asaf Inbal, Akie Mejia, Sarina Patel, Kaylin Rochford, Jennifer Strauss, Jennifer Taggart, Stephen Thompson, and Stephane Zuzlewski in addition to the authors. We thank John Taber and Mladen Dordevik at the IRIS Consortium for guidance on the educational components of MyShake2.0 and the creation of the StationMonitor. This work was funded by the Gordon and Betty Moore Foundation through Grant GBMF 5230 to UC Berkeley.

Publisher's Note Springer Nature remains neutral with regard to jurisdictional claims in published maps and institutional affiliations.

References

Allen, R. (1978). Automatic earthquake recognition and timing from single traces. *BSSA, 68*, 1521–1532.

Allen, R. M. (2012). Transforming earthquake detection? *Science (80–), 335*, 297–298. https://doi.org/10.1126/science.1214650.

Allen, R. M., Cochran, E. S., Huggins, T. J., et al. (2018). Lessons from Mexico's earthquake early warning system. In: Eos (Washington, DC). https://eos.org/features/lessons-from-mexicos-earthquake-early-warning-system. Accessed 18 Jan 2019

Allen, R. M., Cochran, E. S., Huggins, T., et al. (2017). Quake warnings, seismic culture. *Science (80–), 358*, 1111. https://doi.org/10.1126/science.aar4640.

Allen, R. M., Gasparini, P., Kamigaichi, O., et al. (2009). The status of earthquake early warning around the world: An introductory overview. *Seismological Research Letters, 80*, 682–693. https://doi.org/10.1785/gssrl.80.5.682.

Allen, R. M., & Melgar, D. (2019). Earthquake early warning: Advances, scientific challenges, and societal needs. *Annual Review of Earth and Planetary Sciences, 47*, 361–388. https://doi.org/10.1146/annurev-earth-053018-060457.

Atkinson, G. M., & Wald, D. J. (2007). "Did you feel it?" intensity data: A surprisingly good measure of earthquake ground motion. *Seismological Research Letters, 78*, 362–368. https://doi.org/10.1785/gssrl.78.3.362.

Bakun, W. H., & Joyner, W. B. (1984). The ML scale in central California. *Bulletin of the Seismological Society of America, 74*, 1827–1843.

Bossu, R., Roussel, F., Fallou, L., et al. (2018). LastQuake: From rapid information to global seismic risk reduction. *IInternational Journal of Disaster Risk Reductiont, 28,* 32–42. https://doi.org/10.1016/J.IJDRR.2018.02.024.

Brooks, B., Minson, S. E., Böse, M., et al. (2017). Towards internet of things earthquake early warning—a pilot network in Chile. *Seismological Research Letters, 88,* 537. https://doi.org/10.1785/0220170035.

Chung, A. I., Henson, I., & Allen, R. M. (2019). Optimizing earthquake early warning performance: Elarm S-3. *Seismological Research Letters, 90,* 727–743. https://doi.org/10.1785/0220180192.

Chung, A. I., Neighbors, C., Belmonte, A., et al. (2011). The Quake-catcher network rapid aftershock mobilization program following the 2010 M 8.8 Maule, Chile earthquake. *Seismological Research Letters, 82,* 526–532. https://doi.org/10.1785/gssrl.82.4.526.

Clayton, R. W., Heaton, T., Changy, M., et al. (2012). Community seismic network. *Annals of Geophysics, 14,* 54.

Clayton, R. W., Heaton, T., Kohler, M., et al. (2015). Community seismic network: A dense array to sense earthquake strong motion. *Seismological Research Letters, 86,* 1–10. https://doi.org/10.1785/0220150094.

Cochran, E. S., Lawrence, J. F., Christensen, C., & Jakka, R. S. (2009). The Quake–Catcher network: Citizen science expanding seismic horizons. *Seismological Research Letters, 80,* 26–30. https://doi.org/10.1785/gssrl.80.1.26.

Cuellar, A., Espinosa-Aranda, J. M., Suarez, R., et al. (2014). The Mexican Seismic Alert System (SASMEX): Its alert signals, broadcast results and performance during the M 7.4 Punta Maldonado Earthquake of March 20th, 2012. In: Wenzel, F., & Zschau, J. (eds) *Early warning for geological disasters,* pp. 71–87

Earle, P. S. (2010). Earthquake Twitter. *Nature Geoscience, 3,* 221–222.

Ester, M., Kriegel, H. P., Sander, J., & Xu, X. (1996) A density-based algorithm for discovering clusters in large spatial databases with noise. In: *Proceedings of the second international conference on knowledge discovery and data mining,* pp. 226–231

Finazzi, F. (2016). The earthquake network project: Toward a crowdsourced smartphone-based earthquake early warning system. *Bulletin of the Seismological Society of America.* https://doi.org/10.1785/0120150354.

Hoshiba, M. (2014). Review of the nationwide earthquake early warning in Japan during its first five years. In M. Wyss (Ed.), *Earthquake hazard, risk, and disasters* (pp. 505–528). Oxford: Elsevier.

Hsu, T.-Y., Wang, H.-H., Lin, P.-Y., et al. (2016). Performance of the NCREE's on-site warning system during the 5 February 2016 M_w 6.53 Meinong earthquake. *Geophysical Research Letters, 43,* 8954–8959. https://doi.org/10.1002/2016GL069372.

Inbal, A., Kong, Q., Savran, W., & Allen, R. M. (2019). On the feasibility of using the dense MyShake smartphone array for earthquake location. *Seismological Research Letters, 90,* 1209–1218. https://doi.org/10.1785/0220180349.

Kohler, M. D., Cochran, E. S., Given, D., et al. (2017). Earthquake early warning ShakeAlert System: West coast wide production prototype. *Seismological Research Letters.* https://doi.org/10.1785/0220170140.

Kong, Q., Allen, R. M., Kohler, M. D., et al. (2018). Structural health monitoring of buildings using smartphone sensors. *Seismological Research Letters, 89,* 594–602. https://doi.org/10.1785/0220170111.

Kong, Q., Allen, R. M., & Schreier, L. (2016a). MyShake: Initial observations from a global smartphone seismic network. *Geophysical Research Letters.* https://doi.org/10.1002/2016GL070955.

Kong, Q., Allen, R. M., Schreier, L., & Kwon, Y.-W. (2016b). MyShake: A smartphone seismic network for earthquake early warning and beyond. *Science Advances, 2,* e1501055. https://doi.org/10.1126/sciadv.1501055.

Kong, Q., Martin-Short, R., & Allen, R. M. Towards global earthquake early warning with the MyShake smartphone seismic network **(in review)**

Kong, Q., Patel, S., Inbal, A., & Allen, R. M. (2019) Assessing the sensitivity and accuracy of the MyShake smartphone seismic network to detect and characterize earthquakes. https://arxiv.org/abs/1904.09755

Minson, S. E., Brooks, B. A., Glennie, C. L., et al. (2015). Crowdsourced earthquake early warning. *Science Advances.* https://doi.org/10.1126/sciadv.1500036.

Rochford, K., Strauss, J. A., Kong, Q., & Allen, R. M. (2018). MyShake: Using human-centered design methods to promote engagement in a smartphone-based global seismic network. *Frontiers in Earth Science, 6,* 1–14. https://doi.org/10.3389/feart.2018.00237.

Sheen, D., Park, J., Chi, H., et al. (2017). The first stage of an earthquake early warning system in South Korea. *Seismological Research Letters, 88,* 1491–1498. https://doi.org/10.1785/0220170062.

Steed, R. J., Fuenzalida, A., Bossu, R., et al. (2019). Crowdsourcing triggers rapid, reliable earthquake locations.

Strauss, J. A., & Allen, R. M. (2016). Benefits and costs of earthquake early warning. *Seismological Research Letters, 87,* 765–772. https://doi.org/10.1785/0220150149.

Wald, D., Wald, D. J., Quitoriano, V., et al. (2012). USGS "Did You Feel It?" Internet-based macroseismic intensity maps. *Annales Geophysicae.* https://doi.org/10.4401/ag-5354.

Wu, Y.-M., Hsiao, N.-C., Chin, T.-.L, et al. (2013). Earthquake early warning system in Taiwan. In: *Encyclopedia of earthquake engineering.*

Wu, Y., Liang, W., Mittal, H., et al. (2016). Performance of a low-cost earthquake early warning system (P-Alert) during the 2016 M L 6.4 Meinong (Taiwan) Earthquake. *Seismological Research Letters, 87,* 1050–1059. https://doi.org/10.1785/0220160058.

Wu, Y., Mittal, H., Huang, T., et al. (2019). Performance of a low-cost earthquake early warning system (P-Alert) and Shake Map Production during the 2018 Mw 6.4 Hualien, Taiwan, Earthquake. *Seismological Research Letters, 90,* 19–29. https://doi.org/10.1785/0220180170.

(Received June 25, 2019, revised September 25, 2019, accepted September 27, 2019, Published online October 11, 2019)

Pure Appl. Geophys. 177 (2020), 1713–1724
© 2019 Springer Nature Switzerland AG
https://doi.org/10.1007/s00024-019-02344-8

Tsunamis Along the Azores Gibraltar Plate Boundary

M. A. Baptista[1,2] (iD)

Abstract—The Eurasia–Nubia plate boundary between the Azores and the Strait of Gibraltar has been the place of large tsunamigenic earthquakes. The tectonic regime is extensional in the Azores, transcurrent along the Gloria Fault, and compressional in the Strait of Gibraltar. Here, the plate boundary is not clearly defined. The knowledge of past events that occurred in the area constitutes an essential contribution to the evaluations of seismic and tsunami hazard in the North-East Atlantic. In this study, we present an overview of the six major events in the area and show the use of tsunami data to add some constraints on their source. The historical events occurred in the eighteenth-century between 1722 and 1761, while the twentieth-century events occurred between 1941 and 1975. We speculate that major tsunamigenic earthquakes that occur in the Iberia-Maghreb area take place at the boundaries of a lithospheric block approximately defined by the location the six events summarized here, which role and dynamics are not yet understood.

Key words: Tsunamis, Atlantic, Eurasia–Nubia plate boundary.

1. Introduction

The western segment of the Eurasia–Nubia Plate Boundary between the Azores archipelago and the Strait of Gibraltar has been the place of several large tsunamigenic earthquakes.

The tectonic regimes change from an extensional one in the Azores to a compressional one, towards the Strait of Gibraltar. Between 24°W and 20°W, the plate boundary follows a morphological feature, firstly identified by Laughton et al. (1972), called the Gloria Fault (see Fig. 1). This feature is an almost linear structure characterized by right-lateral strike-

slip motion (Buforn et al. 1988, 2004; Argus et al. 1989). Between 20°W and 18°W, it changes strike, from N84E to N71E (Baptista et al. 2016). Large earthquakes related with this plate boundary also occur away from the Gloria Fault (Baptista et al. 2017), suggesting a second source zone 200 km south of Gloria, striking almost East–West (Lynnes and Ruff 1985; Buforn et al. 1988). Buforn et al. (1988) interpreted it as the development of a tectonic micro-block or the re-activation of a previous transform domain (Kaabouben et al. 2008).

The Gloria area has been the origin of several instrumental earthquakes of magnitude higher than 7, followed by small or moderate tsunamis: the 25 November 1941 (Gutenberg and Richter 1949; Udias et al. 1976; Baptista et al. 2016) and the 26 May 1975 (Lynnes and Ruff 1985; Buforn et al. 1988) (see Fig. 1).

To the East of the Gloria Fault, the seismicity is distributed over a large area, making difficult the identification of the plate boundary. Here, the present-day tectonic regime is dominated by the convergence between Africa and Eurasia at ∼ 4 mm/year (Argus and Gordon 1991; DeMets et al. 1994) and the westward migration of the Cadiz Subduction slab ∼ 2 mm/year (Gutscher et al. 2002; Duarte et al. 2013). This area, called the South West Iberian Margin (SWIM) hosts a set of linear and sub-parallel dextral strike-slip faults Zitellini et al. (2009) forming a deformation zone (Miranda et al. 2015). Zitellini et al. (2009) showed that the seismic strain is accommodated by the active morpho-tectonic features and faults already mapped by (Zitellini et al. 2001, 2004; Gracia et al. 2003; Terrinha et al. 2009). This area hosts the sources of the major tsunamigenic earthquakes of the 1 November 1755 (Johnston 1996; Baptista et al. 1998b; Zitellini et al. 2001; Gutscher et al. 2002) and the 31 March 1761 (Baptista et al.

[1] Instituto Superior de Engenharia de Lisboa, Instituto Politécnico de Lisboa, Lisbon, Portugal. E-mail: mavbaptista@gmail.com

[2] Instituto Dom Luiz, Faculdade de Ciências da Universidade de Lisboa, Universidade de Lisboa, Lisbon, Portugal.

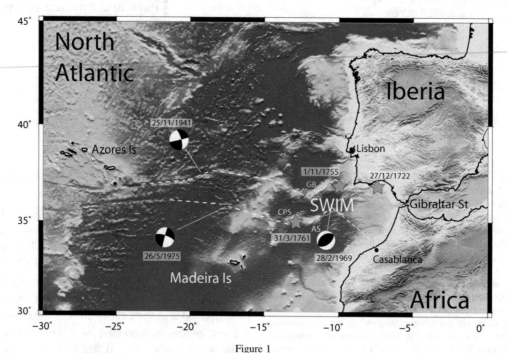

Figure 1
Overview of the study area. Beach balls represent the focal mechanisms of the instrumental events—1941, 1969 and 1975; orange stars represent the presumed epicentres of the historical events 1722, 1755 and 1761. White dashed lines follow main geological lineaments. *GS* Gibraltar Strait, *CP* Coral Patch Seamount, *GB* Gorringe Bank

2006; Wronna et al. (2019). The largest instrumental tsunamigenic earthquake occurred on 28 February 1969 with a magnitude of Ms 8.0, see Fig. 1 for location.

The earthquake and tsunami compilations for the North-East Atlantic area between the Mid Atlantic Ridge in the Azores and the Strait of Gibraltar include reports on the events along this segment of the Eurasia–Nubia Plate Boundary (Baptista and Miranda 2009). According to the Portuguese tsunami catalogue, the oldest historical events occurred in 60 BC and 382 AD, however, paleo-tsunami evidences were found back to 7000 BP (Baptista and Miranda 2009); these events are out of the scope of this study.

The Portuguese tsunami catalogue (Baptista and Miranda 2009) include tsunami or tsunami-like observations following the earthquakes of 27 December 1722, 1 November 1755, and 31 March 1761. The nineteenth century seems to be a quiet period, with no significant events reported. Since the installation of the tide gauge network, three tsunamigenic earthquakes occurred in the twentieth century between the Azores Triple Junction and the

Gulf of Cadiz (see Fig. 1). The first one was the strong strike-slip earthquake of magnitude 8.3 in the Gloria Fault (Baptista et al. 2016) followed by the 28th February 1969 earthquake in the Horseshoe Abyssal plain and the 26 May 1975 earthquake 200 km south of Gloria Fault. This study shows the use of tsunami data to constrain the source of the parent earthquakes with scarce or non-instrumental data available in the western segment of the Eurasia–Nubia plate boundary.

2. Methods

The approach presented here to study tsunami events is based on the compilation of historical and instrumental information on each tsunami to produce datasets of tsunami observations. These datasets include tsunami travel times (TTT), wave heights and run up heights for historical events and tide records for the instrumental ones. The inversion of the tsunami travel times provides information on the source location; The forward simulation of the tsunami using the shallow water approximation supports the selection of a preferred earthquake mechanism that best fits tsunami data. Finally, the inversion of the tsunami waveforms enables the computation of the initial sea surface displacement.

2.1. Backward Ray Tracing of Tsunami Travel Times

The Backward Ray Tracing (BRT) technique consists on the back projection of tsunami wave fronts from each observation point on the coast towards the source area (Miyabe 1934). Gjevik et al. (1997), Baptista et al. (1998b, 2016) present the inversion technique used for the study of all events presented herein. It consists on the definition of dataset of points of interest (POIs) along the coast corresponding to each observation point or one per tide station. The position of each POI is the node of the bathymetric grid closest to the actual location of the observation point or tide station, at a depth not less than 10 m to avoid strong non-linear effects. The back propagation of the wave from each POI uses the algorithm implemented in the Mirone suite (Luis 2007;   http://w3.ualg.pt/~jluis/mirone/main.html). The location of the tsunami source is the minimum of the averaged travel time square errors (in the least squares sense). The spatial distribution of the misfit is given by:

$$\varepsilon^2\,(x,\,y) = \frac{\sum_{i=1}^{n}\left(t_i^{o} - t_i^{p}\right)^2}{n} \qquad (1)$$

where t_i^{o} and t_i^{p} are the observed and predicted travel time for each POI and n is the total number of POIs. The minimum of the averaged travel time square errors is always different from zero because of observation errors, instrumental errors and due to the finite dimension of the fault.

2.2. Forward Tsunami Simulation

In forward tsunami simulations of earthquake induced tsunamis the sea bottom deformation caused by the earthquake is computed using Okada (1985) equations; the ocean bottom deformation is then transferred to the free surface assuming that the water is an incompressible fluid. The initial conditions of the numerical simulation are the free surface

displacement and a null velocity filed. Non-linear shallow water codes are currently used to simulate tsunami propagation from source to coastal areas and inundation phase (tsunami propagation over dry land). The details on the codes used in the six events presented here can be found in (Gjevik et al. 1997; Baptista et al. 1998b, 2003, 2007, 2016; Kaabouben et al. 2008; Wronna et al. 2015, 2019).

The forward simulation of the tsunami using different initial conditions is used to discriminate among the different earthquake mechanisms the one that fits better the tsunami observations.

2.3. Inversion of Tsunami Waveforms

The problem of tsunami waveform inversion was firstly proposed by Satake (1987) leading to the publication of a number of studies focused on the use of inversion methods to estimate the tsunami source were. These studies can be broadly divided in two categories: with a priori assumptions on the fault model Satake (1987, 1993), Johnson et al. (1996), Hirata et al. (2003), Titov et al. (2005) and without a priori assumptions on the fault model and/or the source mechanism Baba et al. (2005), Satake et al. (2005), Tsushima et al. (2009), Wu and Ho (2011), Yasuda and Mase (2012) and Baptista et al. (2016). The results presented herein on the inversion of the source the 25 November 1941—without a priori constraints—are fully described in Baptista et al. (2016). The initial sea surface elevation over the source area was computed using the waveforms recorded at the tide stations shown in Fig. 4 (black dots) by inversion of the following equation

$$\eta_m\,(t_t) = \sum_{i=1}^{nl}\sum_{j=1}^{nc} G_{ij}^{m}(t_t)h_{ij}, \qquad (2)$$

where $G_{ij}^{m}\,(t_t)$ are the empirical Green's functions computed by the shallow water model for t time steps, $(nl \times nc)$ are the set of unit sources and h_{ij} is the amplitude of the initial water displacement attributed to the ij unit source (see Baptista et al. 2016 for details).

3. The Events

3.1. The 27 December 1722 Event

The 27 December 1722 earthquake occurred offshore the south Portuguese coast in the western sector of the Eurasia–Nubia inter-plate domain (see Fig. 1). After the earthquake, tsunami observations were reported in the south coast of Portugal. Baptista et al. (2007) analyzed the historical information on the earthquake and used seismo-tectonic information of the offshore area to propose an offshore source compatible with the flooding described in the historical observations. To set the parameters of the candidate faults, these authors used seismo-stratigraphic analysis of a set of multichannel seismic (MCS) reflection profiles covering the Algarve Basin (Lopes et al. 2006) to infer the strike and size of each candidate source. Each candidate source was used to initiate the tsunami propagation model. A slip of 2 m was assumed to comply with the magnitude of the earthquake; dip and rake angles were fixed with reasonable guesses according to each tectonic style. The results of tsunami forward modelling suggest a source located in the offshore close to 37°01′N, 7°49′W.

Andrade et al. (2016) suggested for a source located in the riverbed. However, Andrade et al. (2016) do not discuss the tectonic implications of this source neither present hydrodynamic simulations supporting the conclusion. Nevertheless, the fact that this is a historical event with scarce observations there is room for alternative candidate sources always close to the shore.

3.2. The 1st November 1755 Event

The Lisbon earthquake of the 1st November 1755, generated a massive transatlantic tsunami that ravaged the coast of SW Portugal, Spain in the Gulf of Cadiz and the Atlantic coast of Morocco. The tsunami reached as far as the Caribbean Islands and UK.

Several authors investigated the source of the Lisbon earthquake, using either macro-seismic data (Solares et al. 1979; Levret 1991), averaged tsunami amplitudes (Abe 1979) of the 28 February 1969, or scale comparisons with the 28 February 1969 event (Johnston 1996). Solares and Arroyo (2004) estimated the magnitude in 8.3 ± 0.4, assuming a source area in the region offshore Iberia close to the epicenter of the 28 February 1969 earthquake (Fukao 1973).

Baptista et al. (1998a, b) presented the first comprehensive analysis of the tsunami observations of this event and used tsunami backward ray tracing techniques to locate the source area between the Gorringe Bank and the southwestern end of the Portuguese coast (see Fig. 2). These authors used a non-linear shallow water code by Mader (1988) to test these sources; the first one corresponds to a source parallel to the Gorringe Bank. The other sources are located in the South West Iberian Margin (SWIM) similar to the elongated shape presented in Fig. 2. All sources were compatible with a magnitude 8.3 parent earthquake. The results show that a source elongated along the SWIM with an azimuth compatible with the results of the backward ray tracing presented in Fig. 2 produce tsunami wave heights and travel times consistent with the observations (Baptista et al. 1998b). Finally, these authors excluded a source compatible with the Gorringe Bank fault because it produces late travel times to the Iberian coast and small tsunami amplitudes. The results presented by Omira et al. (2009) favor this conclusion as the tsunami radiation pattern of a source on the Gorringe Bank shows that most energy is radiated towards NE and Morocco with a minor impact along the Gulf of Cadiz.

The search for the tectonic source of this event was the goal of bathymetric and multi-channel seismic campaigns conducted by Zitellini et al. (1999, 2001), Gutscher et al. (2002) and Gracia et al. (2003). Zitellini et al. (2001) identified several compressive structures located in the SWIM among which the named Marques de Pombal fault (MPf), 100 km offshore southwest of Saint Vicente cape (see Fig. 3). The MPf was interpreted as a possible source of the 1755 earthquake and tsunami. However, hydrodynamic modeling using the MPf with an average slip of 20 m produces wave heights incompatible with the observations along Iberia, Morocco and Madeira archipelago. Moreover, the corresponding seismic moment seems too small to justify the

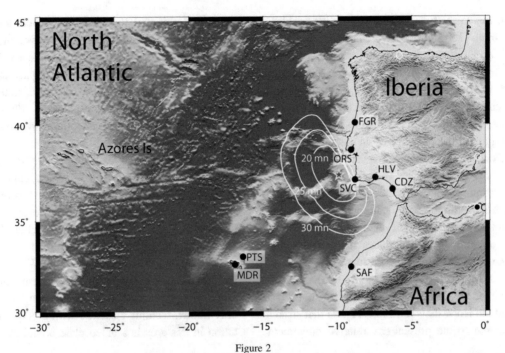

Figure 2

Backward Ray Tracing for the 1755 event. The star coincides with the minimum error of the backward ray tracing simulation. Observation points of the 1755 event used for backward ray tracing simulation. *FGR* Figueira, *ORS* Oeiras, *SVC* Saint Vicent Cape, *HLV* Huelva, *CDZ* Cadiz, *PTS* Porto Santo, *MDR* Madeira, *SAF* Safi (see Baptista et al. 1998b for details)

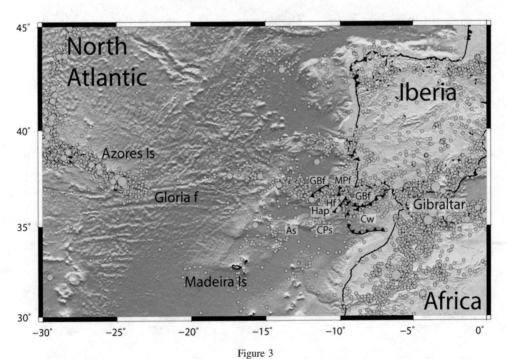

Figure 3

Seismicity of the period 1987–2017 (earthquake magnitudes greater than 3) of the South West Iberia (International Seismological Centre Bulletin) and active structures identified in marine surveys; *GBf* Gorringe Bank, *MPf* Marquês de Pombal fault, *Hf* Horseshoe fault, *GBf* Guadalquivir Bank fault, *Cw* Cadiz Wedge, *As* Ampere seamount, *CPs* Coral Patch seamount, *Hap* Horseshoe abyssal Plain observed macro-seismic field. To solve this problem Baptista et al. (2003) proposed a composite source of Marquês de Pombal and Guadalquivir faults for the source of this event. Figure 3 shows the location of these active structures.

Gutscher et al. (2002) using seismic and tomography images of the crustal structure of the Gulf of Cadiz suggested the presence of an east dipping slab of cold, oceanic lithosphere beneath the Gibraltar–Cadiz wedge (see Fig. 3). The dimensions of this tectonic feature (app. 200 × 200 km) are compatible with an 8.8 magnitude earthquake with a co-seismic displacement of 20 m. However, the tsunami propagation from this source produces late arrivals and small wave heights along the Portuguese west coast and Safi in Morocco (Gutscher et al. 2006).

Gracia et al. (2003) and Zitellini et al. (2004) identified the Horseshoe Fault (see Fig. 4) that strikes perpendicular to the present-day relative movement

between Nubia and Eurasia as a possible source for this earthquake. Omira et al. (2011) investigated the tsunamigenic potential of the Horseshoe and Marques de Pombal faults and discuss the impact along the Gulf of Cadiz and Morocco. However, none of the tsunamis caused by the rupture of these single structures reproduces the set of observations in the area.

Santos et al. (2009) using wave ray analysis relocate the source on the Gorringe bank but the hydrodynamic modelling results are incompatible with the observations producing large wave heights north of Lisbon and small wave heights along the south Portuguese coast and south Spain.

Barkan et al. (2009) proposed a NW–SE trending fault in the Horseshoe Plain almost perpendicular to previously suggested NE–SW trending faults; the only known tectonic structure with a NW–SE orientation in this area is a paleo plate boundary.

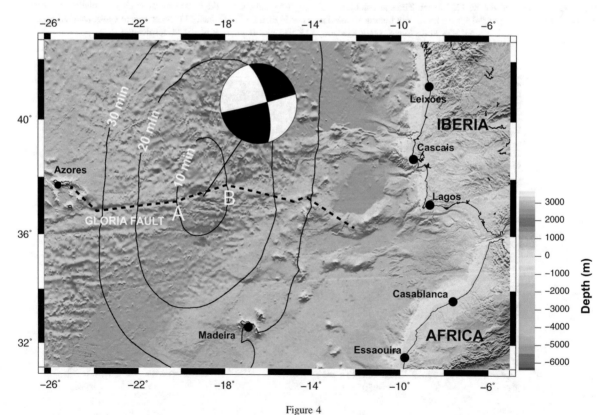

Figure 4

Backward Ray Tracing for the 1941 event. The star coincides with the minimum error of the backward ray tracing simulation. Black dots show the locations of the tide stations used for backward ray tracing simulation and beach ball shows the focal mechanism (see Baptista et al. 2016 for details)

3.3. The 31.03.1761 Event

The 31st March 1761 earthquake and tsunami occurred just 6 years after the Lisbon event in 1755 at 12:05 a.m. Lisbon time. Mezcua and Solares (1983) located the epicenter location was given at 37.00°N, 10.00 W whereas Oliveira (1986) put it at 36.0°N 10.50°W.

The first comprehensive compilation of the earthquake and tsunami is the study by Borlase (1762) in the Philosophical Transactions of the Royal Society of London. These data was recompiled and published by Perrey (1847) and Mallet (1852), Moreira (1984) and De La Torre (1997). Baptista et al. (2006) presented an extensive analysis of macro-seismic data, tsunami data and seaquake observations. The authors used macro-seismic intensity data and epicentral intensity of IX to obtain a new isoseismal map that suggests a source location South West of the Iberian Margin. The use of tsunami travel times in the backward ray tracing simulations lead to a source area compatible with the isoseismal map. Within the limitations of historical data the authors conclude for an earthquake that occurred at 12:01 a.m. on the 31 March 1761 (Lisbon time) with an epicenter located around 34.5°N, 13°W (see Fig. 1). Wronna et al. (2019) proposed a source of 200 km long by 50 km compatible with the geodynamic constraints imposed by Nuvel 1A (DeMets et al. 1994) with a velocity convergence of 3.8 mm/year in the area. The rupture mechanism would be a thrust fault located southwest of the source of the 1755 event close to the Coral Patch seamount (see Fig. 1). These authors proposed a compressive source in this area compatible with the Plate kinematics that reproduces quite well both the near field and the far field tsunami observations.

3.4. The 25 November 1941 Event

The 25 November 1941 earthquake occurred at 18:03:54 UTC. After the earthquake a tsunami was recorded in the tide stations of the North East Atlantic with a maximum amplitude of 40 cm peak to throw in Azores and Madeira islands (Baptista et al. 2016), see Fig. 4 for location of the tide stations, epicenter, and focal mechanism.

Macro-seismic analysis shows largest seismic intensities in the islands of Madeira, Azores and western Portugal VI (MSK). Debrach (1946), Di Filippo (1949) and Moreira (1968) published compilations of macro-seismic data for Morocco, Spain and Portugal respectively.

The epicenter locations given by Antunes (1944) and Gutenberg and Richter (1949) were − 18.9°E, 38.7°N and − 18.5°E, 37.50°N respectively. The study by Baptista et al. (2016) presents a new epicenter location using the phases of the International Seismological Centre Bulletin and readings of nearby stations in Morocco and Spain at − 19.038°E, 37.405°N, and moment magnitude of 8.3. The new epicentre lies west of the epicentre presented by Gutenberg and Richter (1949). Nevertheless, both determinations put the epicentre on the segment AB of the Gloria fault (see Fig. 4).

Baptista et al. (2016) present a comprehensive analysis of the data. The set of tsunami data include records from eight stations available in UK, Azores and Madeira islands, Morocco and three stations in mainland Portugal. The results of the backward ray-tracing simulation locate the source at − 19.00°E, 37.35°N quite close to the epicentre location suggesting that the earthquake's co-seismic deformation induced the tsunami. Baptista et al. (2016) invert the tsunami waveforms without a priori constraints on the focal mechanism using the method proposed by Miranda et al. (2014). The inversion of the tsunami waveforms suggest that the earthquake ruptured approximately 160 km of the plate boundary between 20.249°W and 18.630°W. The inverse problem solution was compatible with the majority of tsunami observations. However, the record in Essaouira (Morocco) (see Fig. 4) could not be explained by an earthquake-induced tsunami. The fact that Debrach (1946) and Rothé (1951) report damages to the submarine cables Brest–Casablanca and Brest–Dakar after the earthquake led Baptista et al. (2016) to speculate that a secondary landslide source caused by the earthquake may explain this observation. The lack of information on the exact location of the submarine cables offshore Morocco and on the location of the rupture prevented the authors to model the landslide.

3.5. The 28 February 1969 Event

On the 28 February 1969, at 2:40 a.m. the strongest earthquake since the 1755 event stroke Portugal, Spain and Morocco. The studies on the earthquake by Udias and Arroyo (1972), Mckenzie (1972), Fukao (1973), Buforn et al. (1988), Grimison (1988) conclude for a thrust mechanism with a small left lateral component.

The tsunami that followed the earthquake reached the first tide stations in the Portuguese coast less than 40 min after the shock (Baptista et al. 1992) being recorded in Spain and Morocco.

Gjevik et al. (1997) computed the location of the tsunami source using tsunami data. The source area, in the Horseshoe Abyssal Plain, encompasses the epicenter of the earthquake but requires the exclusion of some of the tsunami travel times.

The studies on the numerical simulation of the tsunami by Heinrich et al. (1994), Guesmia et al. (1996), Guesmia et al. (1998), Miranda et al. (1996) and Gjevik et al. (1997) used the fault mechanism proposed by Fukao (1973) consisting on thrust with a small strike slip component, N55°E parallel to the Gorringe Bank (see Fig. 3); the fault dimensions being 80 km × 50 km, the dip angle of 52°, average slip on the fault of 3.85 m corresponding to a seismic moment of 6.0×10^{10} Nm and a magnitude of 7.9. These studies could reproduce some of the tsunami observations. However, none of these single source models explain the early arrival with a large downward movement observed in the tide record of Casablanca (Morocco), and some authors suggest the occurrence of a landslide (Gjevik et al. 1997; Pires and Miranda 2001).

3.6. The 26 May 1975 Event

The 26 May 1975, Ms = 7.9 earthquake occurred in the North East Atlantic 200 km south of the Gloria Fault (see Fig. 1) at 17.5°W, 35.9°N. Studies on the focal mechanism by Lynnes and Ruff (1985) "first-motion" solution and Buforn et al. (1988) concluded for a dextral strike-slip event with no significant dip-slip component, compatible with the relative motion between Eurasia and Nubia plates but 200 km south

of the presumed plate boundary. The tsunami was recorded in Portugal and Spain.

Baptista et al. (1992) presents the spectral analysis of the tsunami records of this event. Later, Kaabouben et al. (2008) inverted tsunami travel times to locate the source. These authors selected three candidate sources given by Lynnes and Ruff (1985) "first-motion" solution, Lynnes and Ruff (1985) "Rayleigh Waves" solution and the focal mechanism proposed by Hadley and Kanamori (1975) and propagated the tsunami. The inversion of TTT confirms the epicentre location of 26 May 1975 earthquake close to the Tydeman Fracture Zone almost parallel to the Gloria Fault (see Fig. 1). The shallow water simulations presented by Kaabouben et al. (2008) using Lynnes and Ruff (1985) fault parameters to compute the initial sea surface displacement reproduce well most of the tsunami waveforms of the tide stations in the North East Atlantic (see Kaabouben et al. 2008 for details).

4. Results and Discussion

The results summarized here show how tsunami research contributed to the knowledge of the geological processes along the Azores–Gibraltar segment of the Eurasia–Nubia Africa plate boundary. The information retrieved from historical reports include tsunami travel times, wave heights, first wave polarity and time intervals between primary waves. The inversion of tsunami travel times allows for the computation of an area that includes the tsunami source and the earthquake epicentre for earthquake-induced tsunamis. Inversion of tsunami waveforms provides results on the initial sea surface displacement giving information on the sea bottom deformation caused by the parent earthquake. In spite of the limitations inherent to the use of historical observations and old tide record time series these datasets provide an independent estimate on the location and the extension of the tsunami source and indirectly on the extent of the seismic source.

1. The 27th December 1722 event took place close to the south Portuguese coast and triggered a small tsunami. Hydrodynamic modelling (Baptista et al.

2007) can fairly reproduce the historical reports and is compatible with local geology. This is the easternmost tsunami event known to occur in the Azores-Gibraltar segment of the plate boundary.

2. The inversion of tsunami travel times of the 1 November 1755 places the source of this event in the South West Iberian Margin (SWIM), and exclude a source located in the Gorringe Bank thrust (Baptista et al. 1998b). These results fostered new bathymetric and seismic surveys in the area confirming the existence of active tectonic features that may be the source of the 1755 event (see Fig. 3). All those studies locate the epicentre in the offshore of Iberia. The tectonic structures parallel to the SWIM can reproduce most of the observed tsunami travel times; however, the dimensions of the single structures identified up to now are not compatible with an 8.5–8.75 earthquake and do not reproduce far-field observations. The dimensions of the Cadiz Wedge source (Gutscher et al. 2002) can justify an 8.5 earthquake but fail to reproduce the tsunami travel times on the Portuguese west coast and north of it (Gutscher et al. 2006). The proposed source by Barkan et al. (2009) reproduces part of the observations but does not match any of the neo-tectonic structures identified. The source of this event is still a question of debate, but most of the modelling studies suggest a location close to San Vincent cape with an NNW–SSW strike.

3. The 31 March 1761 event is less well known than the 1755 event, but the re-evaluation of macro-seismic data by De la Torre (1997) together with the Backward Ray Tracing simulations by Baptista et al. (2006) re-locates the source close to Coral Patch seamount. Recently, Wronna et al. (2019) proposed a compressive source in this area compatible with the Plate kinematics that reproduces quite well both the near field and the far field tsunami observations.

4. The study of the source of the 25 November 1941 using tsunami data confirms its location quite close to the epicentre location suggesting that the earthquake's co-seismic deformation induced the tsunami. The inversion of tsunami waveforms made possible a re-evaluation of the tsunami source that encompasses the earthquake epicentre.

The location is compatible with a sudden change of strike close to the eastern end of Gloria Fault (Baptista et al. 2016).

5. All studies concerning the 28 February 1969 tsunami locate the source in the Horseshoe Abyssal Plain. However, the thrust mechanisms proposed until now cannot explain the sizeable downward movement observed in Casablanca. Moreover, active and passive seismic research was never conclusive on the identification of a geo-logical source of this earthquake compatible with the seismic (and tsunami) source.

6. Forward modelling and backward ray tracing for the 26 May 1975 tsunami confirm the seismic location of the event, suggesting that the candidate sources Lynnes and Ruff (1985) deduced from seismic data reproduces first arrival, the polarity of the first movement and first peak amplitudes of the tsunami waves.

The six tsunamigenic events generated in the Azores Gibraltar the plate boundary do not follow a single boundary, as early detected by the studies concerning the 1975 event (Buforn et al. 1988). When plotted together on the seafloor bathymetry (see Fig. 1) they provide an approximate picture of a complex lithospheric block (white dashed line in Fig. 1). We can speculate that major seismic and tsunami events take place at the boundaries of this block, which genesis and dynamics are not yet understood.

Acknowledgements

The author wishes to thank J. M. Miranda for the fruitful discussions and the anonymous reviewers for their comments on the manuscript. The author wishes to thank project TROYO Training of Youth for preparedness Against Marine Hazards.

Publisher's Note Springer Nature remains neutral with regard to jurisdictional claims in published maps and institutional affiliations.

REFERENCES

Abe, K. (1979). Size of great earthquakes of 1837–1974 inferred from tsunami data. *Journal of Geophysical Research: Solid Earth, 84*(B4), 1561–1568.

Andrade, C., Freitas, M. C., Oliveira, M. A., & Costa, P. J. (2016). *On the sedimentological and historical evidences of seismic-triggered tsunamis on the Algarve Coast of Portugal. Plate Boundaries and Natural Hazards* (pp. 219–238). New York: Wiley.

Antunes, M. T. (1944). Notas Acerca do Sismo de 25 de Novembro de 1941. In *4th congress of Portuguese association for the advance of sciences*, Porto, 18–24 June, 1942. Tomo III, 2nd Section, pp. 161–172 (**in Portuguese**).

Argus, D. F., & Gordon, R. G. (1991). No-net-rotation model of current plate velocities incorporating plate motion model NUVEL-1. *Geophysical Research Letters, 18*(11), 2039–2042.

Argus, D. F., Gordon, R. G., DeMets, C., & Stein, S. (1989). Closure of the Africa-Eurasia-North America plate motion circuit and tectonics of the Gloria fault. *Journal of Geophysical Research: Solid Earth, 94*(B5), 5585–5602.

Baba, T., Cummins, R., & Hori, T. (2005). Compound fault rupture during the 2004 off the Kii Peninsula earthquake (M 7.4) inferred from highly resolved coseismic sea-surface deformation. *Earth Planets Space, 57*(3), 167–172.

Baptista, M. A., Heitor, S., Miranda, J. M., Miranda, P., & Victor, L. M. (1998a). The 1755 Lisbon tsunami; evaluation of the tsunami parameters. *Journal of Geodynamics, 25*(1–2), 143–157.

Baptista, M. A., & Miranda, J. M. (2009). Revision of the Portuguese catalog of tsunamis. *Natural Hazards and Earth System Sciences, 9*(1), 25–42.

Baptista, M. A., Miranda, J. M., Batlló, J., Lisboa, F., Luis, J., & Maciá, R. (2016). New study on the 1941 Gloria Fault earthquake and tsunami. *Natural Hazards and Earth System Sciences, 16*(8), 1967–1977.

Baptista, M. A., Miranda, J. M., Chierici, F., & Zitellini, N. (2003). New study of the 1755 earthquake source based on multi-channel seismic survey data and tsunami modeling. *Natural Hazards and Earth System Sciences, 3*(5), 333–340.

Baptista, M. A., Miranda, J. M., Lopes, F. C., & Luis, J. F. (2007). The source of the 1722 Algarve earthquake: Evidence from MCS and Tsunami data. *Journal of Seismology, 11*(4), 371–380.

Baptista, M. A., Miranda, J. M., & Luis, J. F. (2006). In search of the 31 March 1761 earthquake and tsunami source. *Bulletin of the Seismological Society of America, 96*(2), 713–721.

Baptista, M. A., Miranda, J. M., Matias, L., & Omira, R. (2017). Synthetic tsunami waveform catalogs with kinematic constraints. *Natural Hazards and Earth System Sciences, 17*(7), 1253–1265.

Baptista, M. A., Miranda, P. E., & Victor, L. M. (1992). Maximum entropy analysis of Portuguese tsunami data; the tsunamis of 28.02.1969 and 26.05.1975. *Science Tsunami Hazards, 10*(1), 9–20.

Baptista, M. A., Miranda, P. M. A., Miranda, J. M., & Victor, L. M. (1998b). Constrains on the source of the 1755 Lisbon tsunami inferred from numerical modelling of historical data on the source of the 1755 Lisbon tsunami. *Journal of Geodynamics, 25*(1–2), 159–174.

Barkan, R., Uri, S., & Lin, J. (2009). Far field tsunami simulations of the 1755 Lisbon earthquake: Implications for tsunami hazard to the US East Coast and the Caribbean. *Marine Geology, 264*(1–2), 109–122.

Borlase, W. (1762). Some account of the extraordinary agitation of the waters in Mount's-bay, and other places, on the 31st of March 1761: In a letter for the Reverend Dr. C Lyttelton. *Philosophical Transactions of the Royal Society, 52,* 418–431.

Buforn, E., Bezzeghoud, M., Udias, A., & Pro, C. (2004). Seismic sources on the Iberia-African plate boundary and their tectonic implications. *Pure and Applied Geophysics, 161*(3), 623–646.

Buforn, E., Udias, A., & Colombas, M. A. (1988). Seismicity, source mechanisms and tectonics of the Azores-Gibraltar plate boundary. *Tectonophysics, 152*(1), 89–118.

De La Torre, F. R. (1997). Revisión del Catálogo Sísmico Ibérico (años 1760 a 1800) Estudio realizado para Instituto Geográfico Nacional, mediante convenio de investigación número 7.070, de 1997, Madrid.

Debrach, J. (1946). *Raz de marée d'origine sismique enregistré sur le littoral atlantique du Maroc (in French)*. Annales, Maroc: Service de Physique du Globe et de l'institut scientifique Chérifien.

DeMets, C., Gordon, R. G., Argus, D. F., & Stein, S. (1994). Effect of recent revisions to the geomagnetic reversal time scale on estimates of current plate motions. *Geophysical Research Letters, 21*(20), 2191–2194.

Di Filippo, D. (1949). Il terremoto delle Azzore del 25 Nov. 1941. *Annali Geofis, 2,* 400–405.

Duarte, J. C., Rosas, F. M., Terrinha, P., Schellart, W. P., Boutelier, D., Gutscher, M. A., et al. (2013). Are subduction zones invading the Atlantic? Evidence from the southwest Iberia margin. *Geology, 41*(8), 839–842.

Fukao, Y. (1973). Thrust faulting at a lithospheric plate boundary the Portugal earthquake of 1969. *Earth and Planetary Science Letters, 18*(2), 205–216.

Gjevik, B., Pedersen, G., Dybesland, E., Harbitz, C. B., Miranda, P. M. A., Baptista, M. A., et al. (1997). Modeling tsunamis from earthquake sources near Gorringe Bank southwest of Portugal. *Journal of Geophysical Research: Oceans, 102*(C13), 27931–27949.

Gracia, E., Danobeitia, J., Vergés, J., & PARSIFAL Team. (2003). Mapping active faults offshore Portugal (36 N–38 N): Implications for seismic hazard assessment along the southwest Iberian margin. *Geology, 31*(1), 83–86.

Grimison, N. L. (1988). Source mechanisms of four recent earthquakes along the Azores-Gibraltar plate boundary. *Geophysical Journal International, 92,* 391–401.

Guesmia, M., Heinrich, P., & Mariotti, C. (1996). Finite element modelling of the 1969 Portuguese tsunami. *Physics and Chemistry of the Earth, 21*(1–2), 1–6. https://doi.org/10.1016/S0079-1946(97)00001-3.

Guesmia, M., Heinrich, P. H., & Mariotti, C. (1998). Numerical simulation of the 1969 Portuguese tsunami by a finite element method. *Natural Hazards, 17*(1), 31–46.

Gutenberg, B., & Richter, C. F. (1949). *Seismicity of the earth and associated phenomena*. Princeton: Princeton University Press.

Gutscher, M. A., Baptista, M. A., & Miranda, J. M. (2006). The Gibraltar Arc seismogenic zone (part 2): Constraints on a shallow east dipping fault plane source for the 1755 Lisbon earthquake provided by tsunami modeling and seismic intensity. *Tectonophysics, 426*(1–2), 153–166.

Gutscher, M. A., Malod, J., Rehault, J. P., Contrucci, I., Klingelhoefer, F., Mendes-Victor, L., et al. (2002). Evidence for active subduction beneath Gibraltar. *Geology, 30*(12), 1071–1074.

Hadley, D. M., & Kanamori, H. (1975). Seismotectonics of the Eastern Azores-Gibraltar Ridge. In *Transactions of American Geophysical Union* (Vol. 56, No. 12, pp. 1028–1028). Washington, DC: American Geophysical Union.

Heinrich, P., Baptista, M. A., & Miranda, P. (1994). Numerical simulation of the 1969 tsunami along the Portuguese coasts. Preliminary results. *Science of Tsunami Hazards, 12*(1), 3–23.

Hirata, K., Geist, E., Satake, K., Tanioka, Y., & Yamaki, S. (2003). Slip distribution of the 1952 Tokachi-Oki earthquake (M 8.1) along the Kuril trench deduced from tsunami waveform inversion. *Journal of Geophysical Research: Solid Earth (1978–2012), 108*(B4).

Johnson, J. M., Satake, K., Holdahl, S. R., & Sauber, J. (1996). The 1964 Prince William Sound earthquake: Joint inversion of tsunami and geodetic data. *Journal of Geophysical Research: Solid Earth (1978–2012), 101*(B1), 523–532.

Johnston, A. C. (1996). Seismic moment assessment of earthquakes in stable continental regions—III. New Madrid 1811–1812, Charleston 1886 and Lisbon 1755. *Geophysical Journal International, 126*(2), 314–344.

Kaabouben, F., Brahim, A. I., Toto, E., Baptista, M. A., Miranda, J. M., Soares, P., et al. (2008). On the focal mechanism of the 26.05.1975 North Atlantic event contribution from tsunami modeling. *Journal of Seismology, 12*(4), 575–583.

Laughton, A. S., Whitmarsh, R. B., Rusby, J. S. M., Somers, M. L., Revie, J., McCartney, B. S., et al. (1972). A continuous East-West fault on the Azores-Gibraltar Ridge. *Nature, 237,* 217–220. https://doi.org/10.1038/237217a0.

Levret, A. (1991). The effects of the November 1, 1755 "Lisbon" earthquake in Morocco. *Tectonophysics, 193*(1–3), 83–94.

Lopes, F. C., Cunha, P. P., & Le Gall, B. (2006). Cenozoic seismic stratigraphy and tectonic evolution of the Algarve margin (offshore Portugal, southwestern Iberian Peninsula). *Marine Geology, 231*(1–4), 1–36.

Luis, J. F. (2007). Mirone: A multi-purpose tool for exploring grid data. *Computers and Geosciences, 33*(1), 31–41.

Lynnes, C. S., & Ruff, L. J. (1985). Source process and tectonic implications of the great 1975 North Atlantic earthquake. *Geophysical Journal of the Royal Astronomical Society, 82,* 497–510.

Mader, C. L. (1988). *Numerical modeling of water waves.* Berkeley: University of California Press.

Mallet, R. (1852). Report on the facts of earthquake phenomena.

McKenzie, D. (1972). Active tectonics of the Mediterranean region. *Geophysical Journal International, 30*(2), 109–185.

Mezcua, J., & Solares, J. M. M. (1983). Sismicidad del área Iberomogrebí, I.G.N., No. 203, Madrid. http://www.ign.es/web/resources/sismologia/publicaciones//SismicidaddelAreaIbero Mogrebi.pdf.

Miranda, J. M., Baptista, M. A., & Omira, R. (2014). On the use of Green's summation for tsunami waveform estimation: A case study. *Geophysical Journal International, 199*(1), 459–464.

Miranda, J. M., Matias, L., Terrinha, P., Zitellini, N., Baptista, M. A., Chierici, F., Embriaco, D., Marinaro, G., Monna, S., & Pignagnoli, L. (2015). Marine seismogenic-tsunamigenic prone areas: The Gulf of Cadiz. In *Seafloor observatories* (Springer, Berlin, pp. 105–125).

Miranda, J. M., Miranda, P. M. A., Baptista, M. A., & Victor, L. M. (1996). A comparison of the spectral characteristics of observed and simulated tsunamis. *Physics and Chemistry of the Earth, 21*(1–2), 71–74.

Miyabe, N. (1934). An Investigation of the Sanriku Tsunami based on mareogram data. *Bulletin of the Earthquake Research Institute, Tokyo University, Suppl. 1,* 112–126.

Moreira, V. S. (1968). Tsunamis observados em Portugal. Publicacao GEO, 134 **(in Portuguese)**.

Moreira, V. S. (1984). Sismicidade Histórica de Portugal Continental, Separata da Revista do Instituto Nacional de Meteorologia e Geofísica, Portugal, pp. 1–79.

Okada, Y. (1985). Surface deformation due to shear and tensile faults in a half-space. *Bulletin of the Seismological Society of America, 75*(4), 1135–1154.

Oliveira, C. S. (1986). A sismicidade Histórica em Portugal Continental e a Revisão do Catálogo sísmico Nacional, Laboratório Nacional de Engenharia Civil, Proc. 36/1177638, 235, Lisboa.

Omira, R., Baptista, M. A., Matias, L., Miranda, J. M., Catita, C., Carrilho, F., et al. (2009). Design of a sea-level tsunami detection network for the Gulf of Cadiz. *Natural Hazards and Earth System Sciences, 9*(4), 1327–1338.

Omira, R., Baptista, M. A., & Miranda, J. M. (2011). Evaluating tsunami impact on the Gulf of Cadiz coast (Northeast Atlantic). *Pure and Applied Geophysics, 168*(6–7), 1033–1043.

Perrey, A. (1847). Sur les tremblements de terre de la Peninsule Ibérique. Annales des sciences physiques et naturelles, d'agriculture et d'industrie, X. Societé Royale d'agriculture, d'histoire naturelle et des arts utiles, Lyon.

Pires, C., & Miranda, P. (2001). Tsunami waveform inversion by adjoint methods. *Journal of Geophysical Research: Oceans, 106*(C9), 19773–19796.

Rothé, J. P. (1951). The structure of the bed of the Atlantic Ocean. *Eos, Transactions American Geophysical Union, 32*(3), 457–461. https://doi.org/10.1029/TR032i003p00457.

Santos, Â., Koshimura, S., & Imamura, F. (2009). The 1755 Lisbon Tsunami: Tsunami source determination and its validation. *Journal of Disaster Research, 4*(1), 41–52.

Satake, K. (1987). Inversion of tsunami waveforms for the estimation of a fault heterogeneity: Method and numerical experiments. *Journal of Physics of the Earth, 35*(3), 241–254.

Satake, K. (1993). Depth distribution of co-seismic slip along the Nankai Trough, Japan, from joint inversion of geodetic and tsunami data. *Journal of Geophysical Research: Solid Earth (1978–2012), 98*(B3), 4553–4565.

Satake, K., Baba, T., Hirata, K., Iwasaki, S. I., Kato, T., Koshimura, S., et al. (2005). Tsunami source of the 2004 off the Kii Peninsula earthquakes inferred from offshore tsunami and coastal tide gauges. *Earth, Planets and Space, 57*(3), 173–178.

Solares, J. M., & Arroyo, A. L. (2004). The great historical 1755 earthquake. Effects and damage in Spain. *Journal of Seismology, 8*(2), 275–294.

Solares, J. M., Arroyo, A. L., & Mezcua, J. (1979). Isoseismal map of the 1755 Lisbon earthquake obtained from Spanish data. *Tectonophysics, 53*(3–4), 301–313.

Terrinha, P., Matias, L., Vicente, J., Duarte, J., Luis, J., Pinheiro, L., et al. (2009). Morphotectonics and strain partitioning at the Iberia-Africa plate boundary from multibeam and seismic reflection data. *Marine Geology, 267*(3–4), 156–174.

Titov, V. V., Gonzalez, F. I., Bernard, E. N., Eble, M. C., Mofjeld, H. O., Newman, J. C., & Venturato, A. J. (2005). Real-time tsunami forecasting: Challenges and solutions. In *Developing tsunami-resilient communities* (Springer, Amsterdam, pp. 41–58).

Tsushima, H., Hino, R., Fujimoto, H., Tanioka, Y., & Imamura, F. (2009). Near-field tsunami forecasting from cabled ocean bottom pressure data. *Journal of Geophysical Research, 114,* B06309. https://doi.org/10.1029/2008JB005988.

Udias, A., & Arroyo, A. L. (1972). Plate tectonics and the Azores-Gibraltar region. *Nature Physical Science, 237*(74), 67–69.

Udias, A., Arroyo, A. L., & Mezcua, J. (1976). Seismotectonic of the Azores-Alboran region. *Tectonophysics, 31*(3–4), 259–289.

Wronna, M., Baptista, M. A., & Miranda, J. M. (2019). Reanalysis of the 1761 transatlantic tsunami. *Natural Hazards and Earth System Sciences, 19*(2), 337–352.

Wronna, M., Omira, R., & Baptista, M. A. (2015). Deterministic approach for multiple-source tsunami hazard assessment for Sines, Portugal. *Natural Hazards and Earth System Sciences, 15*(11), 2557.

Wu, T. R., & Ho, T. C. (2011). High resolution tsunami inversion for 2010 Chile earthquake. *Natural Hazards and Earth System Science, 11*(12), 3251–3261.

Yasuda, T., & Mase, H. (2012). Real-time tsunami prediction by inversion method using offshore observed GPS buoy data: Nankaido. *Journal of Waterway, Port, Coastal, and Ocean Engineering, 139*(3), 221–231.

Zitellini, N., Chierici, F., Sartori, R., & Torelli, L. (1999). The tectonic source of the 1755 Lisbon Earthquake. *Annali di Geofisica, 42*(1), 49–55.

Zitellini, N., Gràcia, E., Matias, L., Terrinha, P., Abreu, M. A., DeAlteriis, G., et al. (2009). The quest for the Africa-Eurasia plate boundary west of the Strait of Gibraltar. *Earth and Planetary Science Letters, 280*(1–4), 13–50.

Zitellini, N., Mendes, L. A., Cordoba, D., Danobeitia, J., Nicolich, R., Pellis, G., et al. (2001). Source of 1755 Lisbon earthquake, tsunami investigated. *EOS, 82*(26), 285–291.

Zitellini, N., Rovere, M., Terrinha, P., Chierici, F., Matias, L., & Team, B. (2004). Neogene through Quaternary tectonic reactivation of SW Iberian passive margin. *Pure and Applied Geophysics, 161*(3), 565–587.

(Received June 14, 2019, revised September 29, 2019, accepted October 5, 2019, Published online October 17, 2019)

Pure Appl. Geophys. 177 (2020), 1725–1738
© 2019 Springer Nature Switzerland AG
https://doi.org/10.1007/s00024-019-02380-4

Pure and Applied Geophysics

Eurasia–Africa Plate Boundary Affected by a South Atlantic Asthenospheric Channel in the Gulf of Cadiz Region?

M. Catalán,[1] ⓘ Y. M. Martos,[2,3] and J. Martín-Davila[1]

Abstract—The Gulf of Cadiz has been affected by a long and complex geodynamic evolution. The lithospheric structure is poorly understood in this region, and it also shows a diffuse seismicity, spanning over a broad area. The Canary Archipelago has been extensively studied. Nevertheless, there are fundamental topics that are still under debate. No studies have addressed or suggested the possibility of a plausible geodynamic connection between both remarkable locations. In this study we integrate total tectonic subsidence (TTS), Curie point depth (CPD), Bouguer gravity anomaly, and seismic information. TTS shows the existence of a basement bulge in the area of the Canary Archipelago that extends to the north, and in the area of Madeira Island, which extends in a more subtle way to the north, too. Likewise, the CPD reaches the shallowest values in the same location at the Canary Archipelago. These two aspects suggest a cause-effect relationship between TTS and CPD at this specific area. Gravity data and CPD show a linear feature, which links the NW of the Canary Archipelago and the Gulf of Cadiz. The data we manage in this work show remarkable clues as: (a) the absence of a similar signal in the TTS, (b) the fact that CPD it is rather constant along this track, (c) CPD amplitude also almost doubles the values obtained on Canary Archipelago, and (d) the existence of a clear correlation between seismogenic depths and CPD, which points to the existence of a correlation between seismicity and the thermal architecture of the lithosphere. All of these evidences support the presence of a lithospheric thinning between Canary Islands and Gulf of Cádiz area, and in turn, the presence of an asthenospheric channel which feeds and alter locally the Eurasia–Africa Plate Boundary.

Keywords: Bouguer gravity anomaly, Curie depth, total tectonic subsidence, heat flow, thermal anomaly, seismogenic layer, asthenospheric flow, Canary Islands, Gulf of Cadiz.

1. Introduction

The existence of asthenospheric currents was raised for the first time by Alvarez (1982) when he hypothesized the existence of a geographical distribution of mantle outflow different from classical convective cells. It would allow transporting material from areas of lithosphere consumption to areas where it is being created. This hypothesis was first tested by Martos et al. (2014) at the Drake Passage and the Scotia Sea (between South America and the Antarctic Peninsula) based on gravity data, and supported by a recent geothermal heat flow and thermal subsidence study (Martos et al. 2019). Both propose the existence of a transfer of asthenospheric material from the Pacific into the Atlantic where the Shackleton Fracture Zone played an important role in the flow distribution.

The Gulf of Cádiz and the Canary Archipelago are located in the easternmost part of the Atlantic Ocean, and belong to two different plates, Eurasia and Africa plate, respectively. The Azores–Gibraltar Fracture Zone acts as boundary between both plates, until it reaches the Gulf of Cadiz where it becomes what is known as a diffuse plate boundary (Sartori et al. 1994).

The whole area captures a $\sim$ 1700 km belt of volcanism, which is formed by the Canary Islands, the Madeira Archipelago and more than 20 seamounts (Geldmacher et al. 2000). Conducting a study in an area limited by: the Gulf of Cadiz and AGFZ to the north, the Madeira-Tore Rise to the West, the Canary Archipelago in the south, and the African coast to the East, offers an excellent opportunity to gain knowledge about an oceanic margin adjacent to

[1] Geophysics Department, Royal Observatory of the Spanish Navy, San Fernando, 11100 Cádiz, Spain. E-mail: mcatalan@roa.es

[2] Planetary Magnetosphere Laboratory, NASA-Goddard Space Flight Center, Greenbelt, MD, USA.

[3] Department of Astronomy, University of Maryland, College Park, MD, USA.

a continental passive margin, and also an active volcanic environment.

Among the existing hypotheses about the origin of the Canary Archipelago, the most accepted one is the action of a mantle plume (Anguita and Hernán 2000). Same hypothesis can be applied to the Madeira Archipelago too (Morgan 1981). Particularly, several authors support the existence of several hotspot that have been active for 60 Ma under the Jurassic Atlantic oceanic crust off northwest Africa (Holik and Rabinowitz 1991; Roeser et al. 2002, Geldmacher et al. 2000). All of them have a common sublithospheric mantle source (Hoernle et al. 1995).

Our objectives here are to analyse and integrate different geophysical techniques and data including, heat flow, total tectonic subsidence and seismicity to understand the effects of the hotspot track in the area between the Gulf of Cadiz and the Canary Archipelago. Our study supports the influence of the hot spot could have caused a weakening at the base of the lithosphere and in turn, facilitated an asthenospheric connection between the Canary Archipelago and the Gulf of Cadiz.

2. Regional Setting

The relative motion of Africa with respect to Eurasia is defined along the Azores–Gibraltar Fracture Zone (Fig. 1, AGFZ). The AGFZ is a boundary, which extends from the Azores Triple Junction to the Strait of Gibraltar. From a generic approach it behaves like an E–W strike-slip boundary that acts as a separation between the African and Eurasian Plates. According with seismicity this confrontation between Plates evolves from the west to the east from transtension in its western side to transpression in the eastern, with a strike-slip motion in its central part (Jiménez-Munt et al. 2001). Four different geodynamic sectors are recognized: an area of oceanic divergence in the Terceira Ridge area, an intraoceanic transforming zone, an area of oceanic convergence, and finally another of continental convergence (de Vicente et al. 2008). Around 14°W and 13°W there is a change in the tectonic stress regime from extension (to the West) to compression (to the East) (Muñoz-Martín et al. 2010; de Vicente et al.

2008). This change in the tectonic stress regime denotes a change in the Plates boundaries orientation from E–W to NW–SE.

Civiero et al. (2018) analysed a large number of permanent and temporary regional seismic networks deployed in Iberia, Gulf of Cadiz, Morocco and Canary Archipelago. The study provides a detailed image of mantle structures and their tomographic model resolves velocity anomalies. It is not only consistent with previous global tomographic studies but also identifies multiple short wavelength features better. In this work they show their results in terms of P-wave velocity anomalies (δV_p). This magnitude is the departure of P-wave velocities, at a specific depth, from the "average" P-wave velocity for this depth. To obtain the "average" expected P-wave velocity they make use of the velocity model PRISM3D (Arroucau et al. 2017) as a starting point down to the base of the upper-mantle transition zone (MTZ), and the global model LLNL (Simmons et al. 2012) from the base of the MTZ down to a depth of 800 km.

From the spreading magnetic anomalies point of view, the opening of the Atlantic is described by seafloor spreading anomalies belonging to C or M sequences, which lie on both sides of ocean. According with this, the break-up of Africa from America occurred at least 175 Ma ago, while the break-up in the northern Atlantic (between Iberia and North America) was initiated later and then propagated northward. However, the age of the first oceanic spreading magnetic anomalies (M11 and M3) are not well-constrained in the area (Rovere et al. 2004). The study area is dominated by the presence of the J anomaly ridge, a structural ridge which according to Tucholke and Ludwig (1982) lies between M0 and M4, or between M0 and M1 according with Srivastava et al. (2000). It can be easily traced south of the AGFZ and continues north of this fracture zone. Its amplitude, however, decreases considerably northward until it becomes exceedingly small in the vicinity of Galicia Bank. Tucholke and Ludwig (1982) manage two hypotheses for the formation of the J anomaly. One supports that it was formed by excess volcanism at the crest of the Mid-Atlantic Ridge between anomaly M0 and M4. The other origin is related to its passage over a mantle plume due to continental drifting. Material retrieved

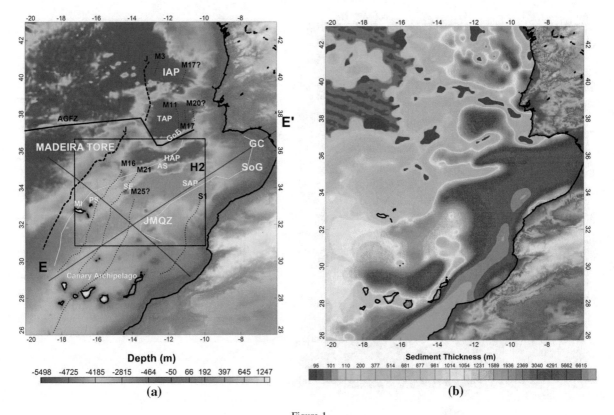

Figure 1
a Bathymetry of the study area. Topography is displayed on land. Thick solid black line denotes the Azores–Gibraltar Fracture Zone (AGFZ). Magnetic lineations are plotted as dotted black lines with their names above. Question mark denotes a doubtful identification. Thick dashed black lines highlight J magnetic anomaly. S1: a linear magnetic anomaly offshore western Africa coast (see text for explanation). *TAP* Tagus Abyssal Plain, *IAP* Iberian Abyssal Plain, *GC* Gulf of Cadiz, *AS* Ampere Seamount, *HAP* Horseshoe Abyssal Plain, *SAP* Seine Abyssal Plain, *SE* Seine Seamount, *GoB* Gorringe Bank, *JMQZ* Jurassic Magnetic Quiet Zone, *SoG* Strait of Gibraltar, *MI* Madeira Island, *PS* Porto Santo Island. Inside two solid thin white lines the map shows a branch of CPD low values that extends from the Canary Archipelago to the Gulf of Cadiz in a continuous way (see Fig. 3). In solid blue line a profile described in Fig. 5. H2 reproduces the location of the P-wave velocity anomaly called in the same way at Civiero et al. (2018). A thick red line (E–E′) reproduces the line called in the same way at Civiero et al. (2018). A box delineates the area studied using spectral methods (Fig. 2b). **b** Sediment thickness map of the study area. Source: Louden et al. (2004)

by drilling at site 1070 in this crust and near anomaly M1 has yielded basement consisting of serpentinized peridotite with only minor occurrences of igneous rocks. This suggests that this crust is not normal oceanic crust, which led Tucholke et al. (2007) to support the second option (mantle plume origin). South of the AGFZ and eastward of the J anomaly other M-series spreading anomalies stand out (e.g. M16, M21 and M25), with being M25 the oldest alignment consistently present in the Eastern Atlantic (Roest et al. 1992). All are distributed along a 300 km-wide-area. The Jurassic Magnetic Quiet Zone (JMQZ) is then extended along a 400 km-wide-

area. This zone ends in a magnetic linear anomaly known as S1 that extends from the northeast of the Canary archipelago parallel to the African west coast to approximately the parallel 33°N (Roeser et al. 2002) (Fig. 1).

North of the AGFZ, anomalies M17 and M20 are located in the area of Tagus Abyssal Plain (TAP), (Srivastava et al. 2000), with M11 being the earliest estimate of the onset of seafloor spreading (Pinheiro et al. 1992). North of the TAP, the situation is not clear, probably as a consequence of this area being a magma-poor margin. This lead Whitmarsh and Miles (1995) to agree with the interpretation of M3 as the

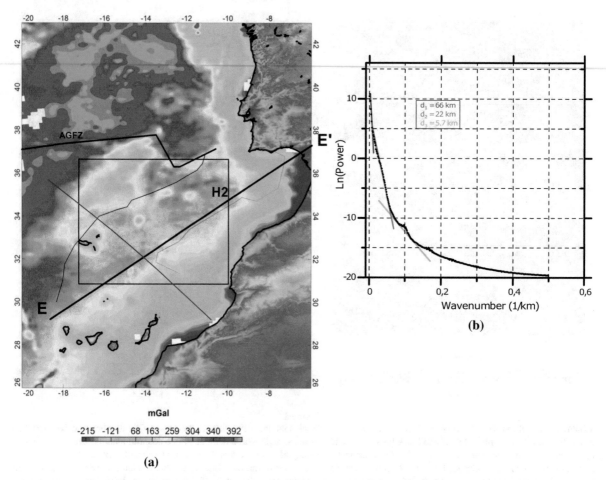

Figure 2
a Complete Bouguer gravity anomaly map. Inside two solid thin black lines the map shows a branch of CPD low values that extends from the Canary Archipelago to the Gulf of Cadiz in a continuous way (see Fig. 3). Topography is displayed on land. A box delineates the area studied using spectral methods. **b** Radial average spectrum of the Complete Bouguer gravity anomalies for the area inside a box. Main horizons are found at 66, 22 and 5.7 km. H2 reproduces the location of the P-wave velocity anomaly called in the same way at Civiero et al. (2018). A thick black line (E–E′) reproduces the line called in the same way at Civiero et al. (2018). In solid blue line a profile described in Fig. 5. A thick solid black line denotes the Azores–Gibraltar Fracture Zone (AGFZ)

oldest anomaly in the Iberia Abyssal Plain (IAP), while Srivastava et al. (2000) identified it as M17.

In the southern area, there are two volcanic provinces: the Madeira and the Canary Archipelagos. Madeira volcanic province forms a nearly 700 km length chain of volcanoes: Madeira Island, Porto Santo Island, Seine and Ampere seamounts. According with its spatial orientation it has been suggested that this group of volcanoes represent a hotspot track (Morgan 1981; Geldmacher et al. 2000).

In the Canary archipelago, there are multiple studies analysing the origin of the Canary Islands. A summary of the state of the art regarding this topic is summarized in the work of Anguita and Hernán (2000).

One aspect that has generated debate is the absence of a geoid anomaly on this group of islands, which are the only ones in the North Atlantic missing this anomaly. Canales and Dañobeitia (1998) used a spectral approach and determined the existence of a swell masked by a thick sedimentary cover. Watts (1994) attributes the lack of this bulging to a high degree of variability in the geophysical properties of rocks in the volcanic environment.

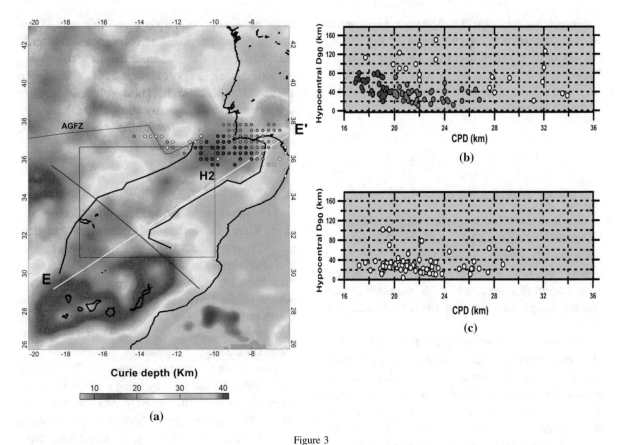

Figure 3

a Curie Point Depth (CPD) of the study area. Inside two solid thick black lines the map shows a branch of low values that extends from the Canary Archipelago to the Gulf of Cadiz in a continuous way. A thin solid black line denotes the Azores–Gibraltar Fracture Zone (AGFZ). Color circles show location of D_{90} values. H2 reproduces the location of the P-wave velocity anomaly called in the same way at Civiero et al. (2018). A thick white line (E–E′) reproduces the line called in the same way at Civiero et al. (2018). In solid blue line a profile described in Fig. 5. **b** Plot showing CPD versus hypocentral D_{90} depths. **c** Plot showing CPD versus hypocentral D_{90} depths once applied filtering criteria (see text). For **a** and **b**: In white circles are displayed D_{90} values with significative dispersion. In orange those D_{90} values that cluster. In red those D_{90} values where the dispersion is minimal

Holik and Rabinowitz (1991) based on the analysis of 6400 km of multichannel seismic data detected a region within the JMQZ that has undergone drastic changes as a result of the passage of the lithosphere over the Canary hotspot. These changes extend through its entire geologic section: seafloor morphology, within sediments, in depth to basement and in deep crustal structure. They considered that this process of lithospheric rejuvenation would have taken place within the domain of the JMQZ off Morocco coast, beginning 60 Ma ago.

Therefore, this thermal disturbance would have affected an area between the Canary Archipelago and approximately the parallel 33°N (as the furthest zone of influence).

In this area Roeser et al. (2002) detects a seismic unit called Unit of Chaotic facies (UCF) within the oceanic crust. It is attributed to igneous activity of the Cenozoic age, and correlates with the thermal anomaly supported by Holik and Rabinowitz (1991). Near the continental slope, the JMQZ ends abruptly at the previously cited magnetic anomaly S1 (Fig. 1). Several questions arise with regard this enigmatic anomaly: Its linear shape, its strong magnetization and uniform polarity over a length of several hundred kilometers, and its abrupt ending northward. Roeser et al. (2002) discussed all these issues, and concluded that according with: (a) the difference in its strike when compared with the lineations in the JMQZ, (b) the observation of Seaward Dipping Reflector Sequence at the S1, and (c) its coincidence with the edge of a salt diapir region supports it to be a marker of the ocean-continent transition area.

3. Data and Methods

3.1. Gravity Data and Procedure

We have used data from the National Center for Environmental Information's database. This database includes 112 cruises. In addition, we have used gravity data from the Spanish Exclusive Economic Zone Project that were obtained in the north Iberian Atlantic Margin between 2001 and 2009 (Druet et al. 2018).

To obtain the Bouguer gravity anomaly, water slab was corrected using a density of 1.03 g/cm^3. Complete Bouguer anomalies were calculated following the Nettleton (1976) procedure. To apply Terrain corrections, we used the SRTM30plusv7grid, 0.5 nautical miles (nm) resolution (approximately 1 km) as local grid, and a subsampled version at 10 km (5.4 nm) of it as a regional grid, which was used for correcting the gravity data beyond 10 km (Becker et al. 2009). We have used in the marine areas a reduction density of 1.64 g/cm^3 as a result of subtracting to 2.67 g/cm^3 a value equal to 1.03 g/cm^3 as the average seawater density (Carbó et al. 2003; Granja-Bruña et al. 2010; Druet et al. 2019). For terrain density we have used 2.67 g/cm^3. Finally, we obtained a Complete Bouguer anomaly grid with 3 nm resolution (approximately 5.5 km) (Fig. 2a).

3.2. Sediment Thickness Data

The thickness of the sediment layer was obtained from the database of Louden et al. (2004) which has 3 nm resolution (approximately 5.5 km). Although other databases exist, such as the National Geophysical Data Center (NGDC) sediment thicknesses database (Straume et al. 2019), or the CRUST1 model (Laske et al. 2013), Louden et al. (2004) provides the highest quality and resolution among those possible in our study area. This compilation was made using digitized contours extracted from 21 published maps produced from seismic and well data (Groupe Galicie 1979; Mauffret et al. 1989; Lefort 1989). The sediment thickness database (Fig. 1b) together with the bathymetry allow us to calculate the basement depth corrected for the effects of isostatic sediment loading.

3.3. Curie Depth

Lithosphere temperature increases with depth and at the Curie Depth the magnetic sources lose their magnetic properties. The Curie Temperature is an intrinsic property of minerals, strongly depending on their crystal structures and composition, and determinant of the behavior of rocks. Hence, the Curie Depth is related to the thermal regime and provides information regarding geodynamic processes and geothermal energy distribution.

Connection between magnetism and Temperature allows inferring the CPD by using a spectral analysis approach on magnetic anomaly data (Martos et al. 2017, 2018, 2019; Arnaiz-Rodriguez and Orihuela 2013; Tanaka et al. 1999). A global Curie point depth (CPD) map with 10 nm resolution (approximately 18 km) was obtained using spectral techniques (based on fractal magnetisation) on magnetic anomalies (Li et al. 2017). We used this dataset (available in their Supplementary material) and retrieve the CPD map for our study area to understand the spatial distribution of this parameter (Fig. 3a).

3.4. Total Tectonic Subsidence

Total tectonic subsidence (TTS) was first introduced by Sawyer (1985) as the difference between the pre-rifting continental crust elevation and the present sediment-unloaded basement depth. This technique assumes that the continental crust was at sea level or near it before extension began. It is a powerful tool to study the evolution of passive continental margins, to infer the amount of crustal extension, or to help the location of crust type boundaries as well as lateral distribution of extension in a basin.

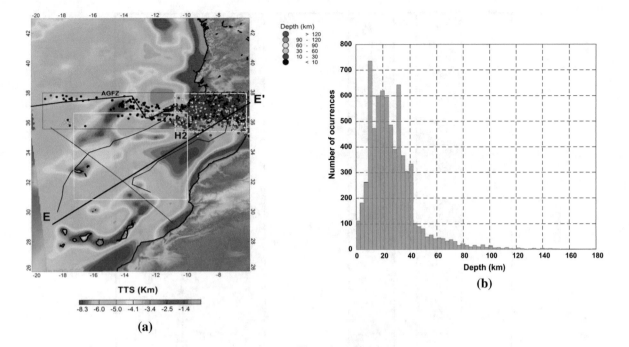

Figure 4
a Total Tectonic Subsidence of the study area. Earthquakes location and hypocentre depth with colour circles from IGN and USGS databases. Topography is displayed on land. Solid black and thick white dashed rectangles delimit the AGFZ and Gulf of Cadiz regions, respectively. A thick solid black line denotes the Azores–Gibraltar Fracture Zone (AGFZ). H2 reproduces the location of the P-wave velocity anomaly called in the same way at Civiero et al. (2018). A thick black line (E–E′) reproduces the line called in the same way at Civiero et al. (2018). A box delineates the area studied using spectral methods on the Complete Bouguer gravity anomaly grid (Fig. 2b). In solid blue line a profile described in Fig. 5. Inside two solid thin black lines the map shows a branch of CPD low values that extends from the Canary Archipelago to the Gulf of Cadiz in a continuous way. **b** Histogram of hypocentral depths corresponding to Gulf of Cadiz

Sawyer (1985) first proposed TTS to identify different types of crust at the US Atlantic margin. Lately it has been used by Henning et al. (2004), and Cunha (2008) to study the evolution of north and central Iberian Atlantic Margin. Heine et al. (2008) used this method to detect and study anomalous tectonic subsidence at large intra-continental basins.

The TTS of the basement is just the sum of the water depth (H), in km, and the sediment thickness (Z), in km, minus the sediment loading effect. To get the water depth (H) we have used the SRTM30PLUS v7 grid (Becker et al. 2009; Smith and Sandwell 1997), and for the sediment thickness we have considered Louden et al. (2004), as it was explained in Sect. 3.2.

There are different existing methods to infer the sediment loading or isostatic corrections (Sykes 1996). Roughly speaking, they are divided depending whether the sediment layer was thicker or thinner than 1 km. Seismic reflection data shows that sediment thickness ranges from 200 m west of 18°W north of AGFZ, to as thick as 7 km all along the African Coast (Louden et al. 2004; Roeser et al. 2002; Holik and Rabinowitz 1991). Such a range of variation makes it difficult to apply suitable algorithms, which do not introduce sharp gradients (artefacts) in the boundaries between thin and thick sediment layer areas. The isostatic correction for thick sediment sequences, where sediment thickness and water depth are known, have been calculated following (Sykes 1996) as:

$$\text{Isos\_Corr} = 0.43\,Z - 0.01\,Z^2 \qquad (1)$$

with *Isos_Corr* being the isostatic correction, and Z the sediment thickness.

This second order polynomial curve which is a simpler version of the one proposed by Sykes (1996) as the line of "best fit" to actual isostatic correction data. This equation does not assume a uniform sediment density, irrespective of either water or

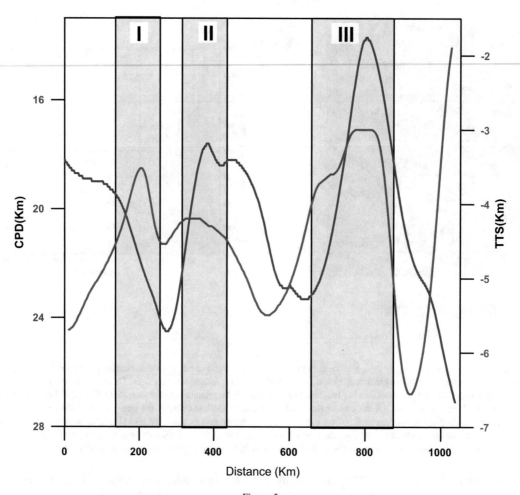

Figure 5
Plot representing CPD (in blue) and TTS (in red) along a profile that crosses the southern part of the study area crossing the Madeira-Tore Rise, north part of Madeira Island, and northern part of Canary Archipelago (in solid blue line for location in Fig. 1). Light grey rectangles (labelled as I, II and III) highlight different segments discussed in the text

depth or total sediment thickness. It was derived by Sykes (1996) using density-depth equations suitable for calcareous, clay and terrigenous sediment sequences. These hypotheses are suitable in our study area (Berger and Rad 1972).

This relationship can be applied on thin sediment layer areas too. Sykes (1996) shows that it is accurate to within 60 m when compared to isostatic corrections calculated using ODP/DSDP data,

Finally, the TTS value is obtained applying (2)[1]:

$$TTS = H - Z + Isos\_Corr \qquad (2)$$

To obtain the TTS map we created a geo-reference database with sediment thickness data at the study area. Using the sediment thickness data and Eq. (1) we easily calculated the sediment loading effect for every geographical location. In the same way we obtained the water depth using the SRTM30PLUS v7 grid. Using Eq. (2) we got the TTS. Finally we interpolated our TTS results at 3 nautical miles resolution (Fig. 4a).

---

[1] Note that water depth is negative, while sediment thickness and isostatic correction are positives.

3.5. Seismicity

We have focused on two specific areas: Gulf of Cadiz and the AGFZ (both delimited with white dashed and with solid black rectangles, respectively, in Fig. 4a). As for the area of the Gulf of Cadiz we have used earthquake locations from the Instituto Geográfico Nacional (IGN) database from 1986 until 2019 (Fig. 4a). As for the AGFZ we have used data from 1973 until 2019 extracted from the USGS catalog. Although the results for the Gulf of Cadiz were similar for IGN and USGS databases, the numbers of items available were higher from the IGN database as there was no lower limit on magnitude (USGS lower limit is 2.5). At AGFZ we have preferred to use data from the USGS database. They not only used data from different partners, but also they manage their own seismographic network of worldwide stations, particularly at Azores Archipelago where they have kept a station active since 1996. This gives us confidence on their hypocentre locations at the AGFZ.

4. Discussion

We have divided this paragraph into two sections: The southern part of the study area, which involves Canary and Madeira Archipelagos, and the second section includes the discussion centered on the northern part of the study area, The AGFZ and the Gulf of Cadiz.

4.1. The Southern Part of the Study Area: Canary and Madeira Archipelagos

The CPD map shows values, which vary from 8 to 40 km. The shallowest values are obtained in oceanic domains, while deeper CPD are located onshore. Particularly CPD values are shallower at the Canary Archipelago. There is a NE–SW CPD signal which progresses from the northwestern part of the Canary Archipelago to the Gulf of Cadiz (delimited by two thick black solid lines in Fig. 3a). Along this feature the CPD amplitude is rather constant (18.5 km). The area along which it runs coincides approximately with the eastern edge of the Madeira Tore Rise. The

feature extends slightly into the western zone of the JMQZ. The lowest CPDs of the study area (8–10 km) are located at the Canary Archipelago, while low CPDs can be found at Gorringe Bank and Ampere Seamount (CPDs of 16.8 km and 17 km, respectively).

The TTS map shows a nearly 250 km-wide area where TTS values are shallow (almost 2.5 km bsl) (Fig. 4a). This area includes the Canary Archipelago and extends northwards. These shallow TTS values show a reasonable geographical correlation with the regional extent of UCF as reported by different authors (Holik and Rabinowitz 1991; Roeser et al. 2002). This region shows the lowest CPD values of the study area (8–10 km).

A TTS high is located on Madeira Island (Fig. 4a). This maximum extends to the north, and it splits into two branches. One of these branches is located over the Madeira-Tore Rise, while the other branch (approximately 640 km-length) addresses towards NE turning E–W in its final section. We have performed a 1038 km profile that cuts the study area from the Atlantic to the African coast (see Fig. 1 for profile location). Throughout this profile we have obtained the TTS and the CPD.

Results obtained are shown in Fig. 5. We have distinguished three zones: Zone I corresponds to the crossing over the Madeira Tore Rise, Zone II corresponds to the area located north of Madeira Island, and Zone III corresponds to the north of the Canary Archipelago.

At Zone I the TTS shows a high, while the CPD shows a clear decrease. This denotes that both variables are uncorrelated. This TTS high is due to the Madeira-Tore Rise. Both TTS and CDP show at Zone II an increase and later decrease with a small lag (approximately 40 km) between both variables. This serves as evidence of correlation between both variables. Along Zone III this correlation is even better as no lag is appreciated between them.

In accordance with all the previous facts we support that these shallow values of TTS can be interpreted as a basement swell caused by thermal processes, probably due to the Madeira (Zone II) and Canary hotspot (Zone III).

Civiero et al. (2018) show, in his Fig. 6, different depth slices from 70 km to 730 km. At the 70 km slice they show a high-velocity feature, which extends from the Canary Archipelago into Western Africa. An additional fast structure progresses from the Canary Archipelago northward, becoming interrupted at 37°N (approximately). There are several areas with intense fast δV_p values (between 0.2 and 0.3 km/s). They are located at the Canary Archipelago, Madeira Island and SW of Cape San Vicente. Civiero et al. (2018) did not discuss those fast structures and focused only on slow bodies.

Their Fig. 7 (profile E–E′) displays a narrow vertical low δV_p value below Canaries, which extends until the MTZ. This anomaly, which reaches lithospheric depth (approximately 100 km), is embedded into a broad fast velocity structure, which extends horizontally through the Atlantic Ocean until parallel 37°N until (approximately Gorringe Bank), reaching shallow depths.

Complete Bouguer anomaly values range from − 200 to 400 mGal. While there is a progressive increase in the Bouguer gravity values from continent to oceanic regime, higher values are obtained east of the Madeira Tore Rise. Lower values are located in the Gibraltar Strait and African coast where we found the thickest sediment layer (larger than 5 km and 7 km, respectively) (Fig. 2a). There is a high broad NE–SW gravity anomaly running subparallel the African coast from northwestern part of Canary Archipelago to the AGFZ. This anomaly is not correlated with any topographic feature. Its amplitude is almost constant at 300 mGal (Fig. 2a). This would correspond to the presence of a dense and deep body, which fits the previous, and cited broad fast velocity structure (Civiero et al. 2018).

In order to estimate the body's centroid location we have obtained radial average spectrum of the Bouguer gravity anomaly map. We have selected a window size large enough to contain the gravity signal of interest. Results are displayed in Fig. 2b. Three main horizons were thus found: horizon 1 is the deepest at (66 ± 10.2) km, horizon 2 at (22 ± 8.6) km and horizon 3 at (5.7 ± 3.2) km. As the study area is in an oceanic domain with an average of 4 km water column in this zone, we interpret horizon 2 as due to lower crust, while horizon 3 is interpreted due to the combined contribution between upper crust and sediment layer.

According with its spectrums' energy content the most prominent body is the deepest one. Based on the crustal and lithospheric thicknesses taken from several studies (Canales and Dañobeitia 1998; Civiero et al. 2019; Fernàndez et al. 2004), horizon 1 is corresponding to the lithosphere. We must point out that spectral methods are estimative and this value provides an approach to the body's centroid location in the area.

Regarding the TTS signal, Fig. 4a does not show a specific feature along this positive Bouguer gravity anomaly. The CPD shows a similar track (inside two solid black lines in Fig. 3a). It is important to highlight that CPD is 19 km and it is rather constant along this track. It also almost doubles the values obtained on the Canary Archipelago, which suggest a non-direct reheating caused by a plume track.

We propose that the high Bouguer anomaly and shallow CPD feature (inside two black solid lines in Fig. 3a) is caused by a lithospheric thinning, which also acts as a large channel for asthenospheric material. We propose three main reasons for this: (a) a thermo-mechanical erosion of the lithospheric mantle probably related to the passage of the mantle plume that originated the Canary Islands, (b) as its trend is sub-parallel to the opening of the Atlantic Ocean, it could have been weakened further the hotspot track and favoured the flow of asthenosphere mantle material north-eastwardly, or (c) a combination of (a) and (b).

4.2. The Northern Part of the Study Area: The AGFZ and Gulf of Cadiz

At the location of the AGFZ, the Bouguer gravity anomaly presents a sharp discontinuity (Fig. 2a) that is probably related to the lithospheric contrast between the African and North Atlantic Plates.

This Bouguer gravity anomaly extends eastward to the Gulf of Cadiz, progressing smoothly towards low values as it reaches continental crust. In addition, a cluster of epicentres ranging from 10 to 30 km depth, is located along the same zone (Fig. 4a).

The branch of low CPD values, which connects the Canary Archipelago to the Gulf of Cadiz, shows a similar image to Bouguer gravity anomaly. A segment of low CPD values follows the AGFZ and extends eastwardly until reaching the Gulf of Cadiz.

Sibson (1982) performed a study in the continental United States away from areas of active subduction. He detected a strong dependence of background microseismicity on the geotherm. He justified it as due to the transition from frictional to quasi-plastic behaviour in quartz-bearing crust. Doser and Kanamori (1986) studied focal depths corresponding to 1000 relocated earthquakes in the Imperial Valley—southern Peninsular Ranges (Southern California, USA). This study revealed that the deepest seismicity in the study area is associated with the regions of lowest heat flow. Tanaka and Ishikawa (2005) conducted a study analysing the thickness of the layer (D_{90}), defined as the depth above which 90% of earthquakes occur. They found good correlations between heat flow and D_{90}, and concluded that Curie point depth is a useful indicator of the crustal thermal structure beneath the Japanese islands.

Figure 4b shows a histogram of hypocentral depths corresponding to the Gulf of Cadiz (delimited by a white dashed rectangle in Fig. 4a). According to the histogram the average hypocentral depth is 25 km. The histogram also shows that 90% of the depths are less than 42 km, while 99% of the depths are located in the upper 78 km. This means that the area of the Gulf of Cadiz shows a drastic decrease in the number of earthquakes from a depth threshold equal to 42 km being absent for depths larger than 78 km. Among the 63 hypocentral depths larger than 78 km only two in those 63 present magnitudes larger than 4 (4.4 and 4.5, respectively).

We have obtained the D_{90} depth in the AGFZ and Gulf of Cadiz joined geographical frame in points separated every 0.3° by taking into account the number of seismic events within a distance of 0.1° (see Fig. 3a).

Figure 3b represents D_{90} versus CPD for the whole area. Some dispersion can be observed which decreases at smaller CPD. It is possible to recognize a cluster that starts from a CPD of 27 km, and extends towards shallower CPD values (in orange and red circles in Fig. 3b).

If we locate these values (in orange and red circles in Fig. 3a), a large number of them are located in the western zone of the Gulf of Cádiz, and on the area where the CPD reaches its lowest values.

Within this segment, another area where dispersion is minimal is identified (CPD < 19.8 km). This cluster has an average D_{90} equal to 53 km. There is only one D_{90} value (113 km) in that segment (CPD < 19.8 km), which does not fit its surroundings. It corresponds to an earthquake located in the NE area of the Horseshoe Abyssal Plain. Its magnitude was 2.8. The Instituto Potuguês do Mar e da Atmosfera's solution locates it at a depth of 10 km.

Those locations where CPD is shallower (red circles in Fig. 3a) fall in the oceanic domain, specifically at the NE end of the shallow CPDs linear structure that extends from the Canary Archipelago to the area located west of Cape San Vicente (delimited by thin black solid lines in Fig. 3a). The fact that dispersion in the D_{90} vs CPD graph is minimal suggests a correlation between the D_{90} and the thermal structure of the lithosphere (represented by the CPD) (in red circles in Fig. 3b).

We have not made any specific earthquakes selection to obtain D_{90} (i.e.: based on the accuracy of their locations). This was motivated by the interest of having as many samples as possible in the study region. To evaluate if this could have conditioned our solution, we have made an additional analysis. We carefully selected well-determined earthquakes with standard errors for horizontal and vertical coordinates of less than 10 km, and we only accept those that occurred after 1997.

As our analysis is based on detecting clusters in a 20 km vertical scale (Fig. 3b), a 10 km error threshold is considered acceptable. Year 1997 was chosen as it was in 1996 when broadband stations started being deployed at SE Iberia.

Figure 3c shows a distribution similar to that one obtained when no selection was applied (Fig. 3b). Particularly, Fig. 3c does not show scattered values for CPD shallower than 19.8 km, which means that this CPD value still represents a limit.

In addition, points to the left of 19.8 km (CPD) are located in the same geographical area occupied by the red dots in Fig. 3b, and vice versa, none of those points located in that geographical area (in red in Fig. 3a) are located right of 19.8 km (CPD) (Fig. 3b, c).

Summarizing, (1) we consider that the results shown in Fig. 3b are valid, (2) CPD equal to 19.8 km acts as a limit where for smaller values of D_{90}, scattered values disappeared, and (3) they are located on a limited geographical location.

It leads us to conclude the existence of a correlation between the CPD (or heat flow) and the maximum depth at which earthquakes can occur in this area.

The V_p seismic tomography (Civiero et al. 2018) shows a ≈ 350 km blob-like structure (label as H2 in their Fig. 7e) west of the Strait of Gibraltar and below the Gorringe Bank. It dips slightly eastward to Cape San Vicente down to a depth of at least 300 km. The magnitude of the P-wave velocity anomaly at H2 decreases from $\delta V_p \approx 0.2$ to 0.1 km/s above the MTZ indicating the presence of a dense body in the so-called H2 that corresponds to the feature labelled in the same way (H2) in the Figs. 1, 2, 3 and 4. This feature also is characterized by a high in the Bouguer gravity anomaly map in agreement with the existence of a higher density area. According with seismic tomography, gravity, and CPD information we propose a lithospheric thinning in the San Vicente area. This lithospheric thinning is also supported by Fernàndez et al. (2004). The decrease in seismicity from 78 km depth onward, or the well-defined 53 km average D_{90} obtained for CPDs smaller than 19.8 km could point to a change in rheology. Based on the CPD pattern, the asthenospheric channel running from the Canary Archipelago towards the Gulf of Cadiz area also penetrates the North Atlantic Plate through the Tagus Abyssal Plain.

5. Conclusions

Here, we have integrated TTS, CPD, Bouguer gravity anomaly, and seismicity information. TTS and CPD put in evidence the existence of a basement bulge in the area of the Canary Archipelago that extends to the north, and in the area of Madeira Island, which extends in a more subtle way to the north, too. Gravity data and CPD show a linear feature, which connects the NW of the Canary Archipelago and the Gulf of Cadiz. The absence of a similar signal in the TTS, the fact that CPD it is rather constant along this track, and that CPD amplitude also almost doubles the values obtained on Canary Archipelago, supports the presence of an asthenospheric connection between both locations. Our work also shows the existence of a clear correlation between seismogenic depth (D_{90}) and CPD in the easternmost part of Gulf of Cadiz area. The diffuse pattern of the seismicity in this region, the existence of a high CPD, and the previous cited correlation between CPD and D_{90} indicate that it is likely that the present plate boundary is affected thermally and in turn, mechanically, due to the influence of the underlying asthenospheric flow created possibly by the effect of the plume track. In addition, and based on the CPD pattern we support that this area serves as connection of asthenospheric material from the African plate into the North Atlantic plate, running eastward of the Tagus Abyssal Plain affecting the deformation and geodynamic activity of the Eurasia–Africa Plate boundary.

Acknowledgements

This study has been funded through project RTI2018-099615-B-I00.

Compliance with Ethical Standards

Conflict of interest The authors declare that they have no conflict of interest.

Publisher's Note Springer Nature remains neutral with regard to jurisdictional claims in published maps and institutional affiliations.

REFERENCES

Alvarez, W. (1982). Geological evidence for the geographical pattern of mantle return flow and the driving mechanism of plate tectonics. *Journal of Geophysical Research, 87*(B8), 6697–6710.

Anguita, F., & Hernán, F. (2000). The Canary Islands origin: a unifying model. *Journal of Volcanology and Geothermal Research, 103*(1–4), 1–26.

Arnaiz-Rodriguez, M., & Orihuela, N. (2013). Curie point depth in Venezuela and Eastern Caribbean. *Tectonophysics, 590*(2013), 38–51.

Arroucau, P., Custódio, S., Civiero, C., Dias, N. A., & Silveira, G. (2017). PRISM3D: a preliminary 3D reference seismic model of the crust and upper mantle beneath Iberia. *General Assembly Conference Abstracts, 19,* EGU2017-16801.

Becker, J. J., Sandwell, D. T., Smith, W. H. F., Braud, J., Binder, B., Depner, J., et al. (2009). Global bathymetry and elevation data at 30 arc seconds resolution: SRTM30_PLUS. *Marine Geodesy, 32*(4), 355–371.

Berger, W. H., & Rad, U. (1972). Cretaceous and Cenozoic sediments from the Atlantic Ocean. In D. E. Hayes & A. C. Pimm (Eds.), *Initial reports, deep sea drilling project* (Vol. 14, pp. 787–954). Washington: U.S. Government Pinting Office.

Canales, J. P., & Dañobeitia, J. (1998). The Canary Islands swell: a coherence analysis of bathymetry and gravity. *Geophysical Journal International, 132,* 479–488.

Carbó, A., Muñoz-Martín, A., Llanes, P., Alvarez, J., & EEZ Working Group. (2003). Gravity analysis offshore the Canary Islands from a systematic survey. *Marine Geophysical Researches, 24*(1–2), 113–127.

Civiero, C., Custódio, S., Rawlinson, N., Strak, V., Silveira, G., Arroucau, P., et al. (2019). Thermal nature of mantle upwellings below the Ibero-Western maghreb region inferred from teleseismic tomography. *of Geophysical Research: Solid Earth.* https://doi.org/10.1029/2018JB016531.

Civiero, C., Strak, V., Custódio, S., Silveira, G., Rawlinson, N., Arroucau, P., et al. (2018). A common deep source for upper-mantle upwellings below the Ibero-western Maghreb region from teleseismic P-wave travel-time tomography. *Earth and Planetary Science Letters, 499,* 157–172.

Cunha, T. (2008). Gravity anomalies, Flexure and the Thermo-Mechanical evolution of the West Iberia Margin and its conjugate of Newfounland. PhD. Thesis. Oxford University.

de Vicente, G., Cloetingh, S., Muñoz-Martín, A., Olaiz, A., Stich, D., Vegas, R., et al. (2008). Inversion of moment tensor focal mechanism for active stresses around the microcontinent Iberia: Tectonic implications. *Tectonics, 27,* TC1009. https://doi.org/10.1029/2006tc002093.

Doser, D. I., & Kanamori, H. (1986). Depth of seismicity in the Imperial Valley region (1977–1983) and its relationship to heat flow, crustal structure and the October 15, 1979, earthquake. *Journal of Geophysical Research, 91*(B1), 675–688.

Druet, M., Muñoz-Martín, A., Granja-Bruña, J. L., Carbó-Gorosabel, A., Acosta, J., Llanes, P., et al. (2018). Crustal structure and continent-ocean boundary along the Galicia continental margin (NW Iberia): Insights from combined gravity and seismic interpretation. *Tectonics.* https://doi.org/10.1029/2017TC004903.

Druet, M., Muñoz-Martín, A., Granja-Bruña, J. L., Carbó-Gorosabel, A., Llanes, P., Catalán, M., et al. (2019). Bouguer anomalies of the NW Iberian continental margin and the adjacent abyssal plains. *Journal of Maps, 15*(2), 635–641.

Fernàndez, M., Marzan, I., & Torne, M. (2004). Lithospheric transition from the Variscan Iberian Massif to the Jurassic oceanic crust of the Central Atlantic. *Tectonophysics, 386*(1–2), 97–115.

Geldmacher, J., van den Bogaard, P., Hoernle, K., & Schmincke, H.-U. (2000). The $^{40}Ar/^{39}Ar$ age dating of the Madeira Archipelago and hotspot track (eastern North Atlantic). *Geochemistry, Geophysics, Geosystems, 1,* 1008. https://doi.org/10.1029/1999GC000018.

Granja-Bruña, J. L., Muñoz-Martín, A., Ten-Brink, U. S., Carbó-Gorosabel, A., Llanes Estrada, P., Martín Davila, J., et al. (2010). Gravity modeling of the Muertos Trough and tectonic implications (north-eastern Caribbean). *Marine Geophysical Researches, 31*(4), 263–283.

Groupe Galicie (1979). The continental margin of Galicia and Portugal, acoustic stratigraphy, dredge stratigraphy and structural evolution. In Ryan, W. B. F., & Sibuet, J. C. (editors), Procedings of the deep sea drilling project, Leg 47, US Government Printing Office, Washington, DC, pp 633–662.

Heine, C., Müller, R. D., Steinberger, B., & Torsvik, T. H. (2008). Subsidence in intracontinental basins due to dynamic topography. *Physics of the earth and Planetary Interiors.* https://doi.org/10.1016/j.pepi.2008.05.008.

Henning, A. T., Sawyer, D. S., & Templeton, D. C. (2004). Exhumed upper mantle within the ocean-continent transition on the northern West Iberia margin: Evidence from prestack depth migration and total tectonic subsidence analyses. *Journal of Geophysical Research.* https://doi.org/10.1029/2003JB002526.

Hoernle, K., Zhang, Y.-S., & Graham, D. (1995). Seismic and geochemical evidence for large-scale mantle upwelling beneath the eastern Atlantic and western and central Europe. *Nature, 374,* 34–39.

Holik, J. S., & Rabinowitz, P. D. (1991). Effects of Canary hotspot volcanism on structure of oceanic crust off Morocco. *Journal of Geophysical Research.* https://doi.org/10.1029/2003JB002526.

Jiménez-Munt, I., Fernàndez, M., Torné, M., & Bird, P. (2001). The transition from linear to diffuse plate boundary in the Azores–Gibraltar region: results from a thin-sheet model. *Earth and Planetary Science Letters, 192*(2), 175–189.

Laske, G., Masters, G., Ma, Z., & Pasyanos, M. (2013). Update on CRUST1.0—A 1-degree global model of earth's crust. *General Assembly Conference Abstracts, 15,* Abstract EGU2013-2658.

Lefort, J. P. (1989). *Basement correlation across the North Atlantic.* New York: Springer.

Li, C.-F., Lu, Y., & Wang, J. (2017). A global reference model of Curie-point depths based on EMAG2. *Nature Publishing Group.* https://doi.org/10.1038/srep45129.

Louden, K. E., Tucholke, B. E., & Oakey, G. N. (2004). Regional anomalies of sediment thickness, basement depth and isostatic crustal thickness in the North Atlantic Ocean. *Earth and Planetary Science Letters.* https://doi.org/10.1016/j.epsl.2004.05.002.

Martos, Y. M., Catalán, M., & Galindo-Zaldívar, J. (2019). Curie depth, heat flux and thermal subsidence studies reveal the Pacific mantle outflow through the Scotia Sea. *Journal of Geophysical Research: Solid Earth.* https://doi.org/10.1029/2019JB017677.

Martos, Y. M., Catalán, M., Jordan, T. A., Golynsky, A., Golynsky, D., Eagles, G., et al. (2017). Heat flux distribution of Antarctica unveiled. *Geophysical Research Letters, 44*(22), 11–417.

Martos, Y. M., Galindo-Zaldívar, J., Catalán, M., Bohoyo, F., & Maldonado, A. (2014). Asthenospheric Pacific-Atlantic flow barriers and the West Scotia Ridge extinction. *Geophysical Research Letters.* https://doi.org/10.1002/2013GL058885.

Martos, Y. M., Jordan, T. A., Catalán, M., Jordan, T. M., Bamber, J. L., & Vaughan, D. G. (2018). Geothermal heat flux reveals the Iceland hotspot track underneath Greenland. *Geophysical Research Letters, 45*(16), 8214–8222.

Mauffret, A., Mougenot, D., Miles, P. R., & Malod, J. A. (1989). Results from multi-channel reflection profiling of the Tagus Abyssal Plain (Portugal)-Comparison with the Canadian Margin. In Tankard, A. J., & Balkwill, H. R. (editors), *Extensional Tectonics and Stratigraphy of the North Atlantic Margins.* American Association of Petroleum Geologist Memoir, 46, 379-393.

Morgan, W. J. (1981). Hotspot tracks and the opening of the Atlantic and Indian Oceans. In C. Emiliani (Ed.), *The sea: Oceanic lithosphere* (Vol. 7, pp. 443–487). New York: Wiley.

Muñoz-Martín, A., de Vicente, G., Fernández-Lozano, J., Cloetingh, S., Willingshofer, E., Sokoutis, D., et al. (2010). Spectral analysis of the gravity and elevation along the western Africa-Eurasia plate tectonic limit: Continental versus oceanic lithospheric folding signals. *Tectonophysics, 495*(3–4), 298–314.

Nettleton, L. L. (1976). *Gravity and magnetic in oil exploration*. New York: Mac Graw-Hill.

Pinheiro, L. M., Whitmarsh, R., & Miles, P. (1992). Ocean-continent boundary off the western continental margin of Iberia-11. Crustal structure in the Tagus Abyssal Plain. *Geophysical Journal International, 109*, 106–124.

Roeser, H. A., Steiner, C., Schreckenberger, B., & Block, M. (2002). Structural development of the Jurassic Magnetic Quiet Zone off Morocco and identification of Middle Jurassic magnetic lineations. *Journal of Geophysical Research.* https://doi.org/10.1029/2000JB000094.

Roest, W. R., Danobeitia, J. J., Verhoef, J., & Collette, B. J. (1992). Magnetic anomalies in the canary basin and the Mesozoic evolution of the central North Atlantic. *Marine Geophysical Researches, 14*(1), 1–24.

Rovere, M., Ranero, C., Sartori, R., & Torelli, L. (2004). Seismic images and magnetic signature of the Late Jurassic to Early Cretaceous Africa-Eurasia plate boundary off SW Iberia. *Geophysical Journal International, 158*, 554–568.

Sartori, R., Torelli, L., Zitellini, N., Peis, D., & Lodolo, E. (1994). Eastern segment of the Azores–Gibraltar line (central-eastern Atlantic): an oceanic plate boundary with diffuse compressional deformation. *Geology, 22*, 555–558.

Sawyer, D. S. (1985). Total tectonic subsidence: A parameter for distinguishing crust type at the US Atlantic continental margin. *Journal of Geophysical Research, 90*(B9), 7751–7769.

Sibson, R. H. (1982). Fault zone models, heat flow, and the depth distribution of earthquakes in the continental crust of the United States. *Bulletin of the Seismological Society of America, 72*(1), 151–163.

Simmons, N. A., Myers, S. C., Johannesson, G., & Matzel, E. (2012). LLNL-G3Dv3: global P wave tomography model for improved regional and teleseismic travel time prediction. *Journal of Geophysical Research: Solid Earth.* https://doi.org/10.1029/2012JB009525.

Smith, W. H. F., & Sandwell, D. T. (1997). Global Sea floor topography from satellite altimetry and Ship Depth soundings. *Science, 277*, 1956–1962.

Srivastava, S., Sibuet, J., Cande, S., Roest, W., & Reid, I. D. (2000). Magnetic evidence for slow seafloor spreading during the formation of the Newfoundland and Iberian margins. *Earth and Planetary Science Letters, 182*, 61–76.

Straume, E. O., Gaina, C., Medvedev, S., Hochmuth, K., Gohl, K., Whittaker, J. M., et al. (2019). GlobSed: updated total sediment thickness in the world's oceans. *Geochemistry, Geophysics, Geosystems.* https://doi.org/10.1029/2018GC008115.

Sykes, T. J. S. (1996). A correction for sediment load upon the ocean floor: Uniform versus varying sediment density estimations—implications for isostatic correction. *Marine Geology, 133*(1), 35–49.

Tanaka, A., & Ishikawa, Y. (2005). Crustal thermal regime inferred from magnetic anomaly data and its relationship to seismogenic layer thickness: The Japanese islands case study. *Physics of the Earth and Planetary Interiors.* https://doi.org/10.1016/j.pepi.2005.04.011.

Tanaka, A., Okubo, Y., & Matsubayashi, O. (1999). Curie point depth based on spectrum analysis of the magnetic anomaly data in East and Southeast Asia. *Tectonophysics, 306*, 461–470.

Tucholke, B. E., & Ludwig, W. J. (1982). Structure and origin of the J Anomaly Ridge, western North Atlantic Ocean. *Journal of Geophysical Research, 87*(B11), 9389–9407.

Tucholke, B.E., Sawyer, D.S., Sibuet, J.-C. (2007). Breakup of the Newfoundland-Iberia rift. In Karner, G.D., Manatschal, G. & Pinheiro, L.M. (editors) Imaging, mapping and modelling continental lithosphere extension and breakup. Geological Society, London, Special Publications, vol. 282, pp 9–46, http://dx.doi.org/10.1144/SP282.2.

Watts, A. B. (1994). Crustal structure, gravity anomalies and flexure of the lithosphere in the vicinity of the Canary Islands. *Geophysical Journal International, 119*(2), 648–666.

Whitmarsh, R., & Miles, P. (1995). Models of the development of the West Iberia rifted continental margin at 40°30′ N deduced from surface and deep-tow magnetic anomalies. *Journal of Geophysical Research, 100*(B3), 3789–3806.

(Received June 26, 2019, revised November 22, 2019, accepted November 25, 2019, Published online December 4, 2019)

Pure Appl. Geophys. 177 (2020), 1739–1745
© 2019 Springer Nature Switzerland AG
https://doi.org/10.1007/s00024-019-02323-z

Large Earthquakes and Tsunamis at Saint Vincent Cape before the Lisbon 1755 Earthquake: A Historical Review

AGUSTÍN UDÍAS[1]

Abstract—The occurrence of large earthquakes followed by tsunamis west off Saint Vincent Cape before the 1755 Lisbon earthquakes is the subject of a historical review. There are historical evidences of the occurrence of such earthquakes. Among the possible earthquakes with some historical records in this area, those with better information are those at years 241 B.C., 216 B.C., 881, 1356 and 1531. Although not a subduction zone very large earthquakes have happened in this area.

Key words: Large earthquakes, tsunamis, Saint Vincent Cape, Lisbon earthquake.

1. Introduction

In the 1st of November of 1755, a very large earthquake, known as the Lisbon earthquake, caused many casualties and large damage in this city and in the southwest part of the Iberian Peninsula and northwest of Morocco. The earthquake with epicenter offshore the Saint Vincent Cape generated a large tsunami which devastated the coasts of the region, contributing to damages and casualties and was detected across the Atlantic Ocean in the Caribbean Islands. The earthquake was felt in the whole Iberian Peninsula, large parts of Morocco, very slightly in France near the Pyrenees and produced ripples on the lakes in Switzerland. Tectonically the earthquake was located at the eastern end of the Azores-Gibraltar fault, the westernmost part of the boundary between the plates of Eurasia and Africa (Buforn and Udías 2010). Its magnitude could have reached nine degrees, similar to the recent very large earthquakes (M > 9), followed by tsunamis, which have occurred

in Sumatra in 26 December 2004 (M_w = 9.2) and in Tohoku, Japan, in 11 March 2011 (M_w = 9.0), with devastating effects in casualties and damages. The seismicity of the region (Fig. 1) shows a continuous occurrence of earthquakes, the largest among them that of 28 February 1969, (M_s = 8), which also produced a small tsunami. Earthquakes followed by tsunamis have also occurred in other locations of the Ibero-Maghrebian region east of the Strait of Gibraltar, such as those of Oran (1790) and Boumerdes (2003), but here we will only consider those in the Saint Vincent Cape region.

It is pertinent, then, to ask if there is historical information about the occurrence of other very large earthquakes followed by tsunamis in this region, before that of the Lisbon earthquake of 1755. In the Tohoku region, for example, a similar earthquake as that of 2011, followed also by a large tsunami, occurred in year 869 (Dunbar et al. 2011). This shows the importance of historical studies about the occurrence of large earthquakes and tsunamis previous to a given one, even in the distant past, like in the case of the Lisbon earthquake. In this regard, special interest have the studies of historical earthquakes that have occurred in the Iberian Peninsula, such as those by Udías (2015, 2017) and Álvarez-Martí-Aguilar (2017) and the geological studies of the coastal region of Cadiz, such as those by Campos (1991, 1992) and Alonso et al. (2015).

Fort the study of antique earthquakes (before year 500 A.D.) in the Saint Vincent Cape region, information should be looked for in Greek and Roman historians. Unfortunately, nothing is found in these historians about the occurrence of large earthquakes in the Iberian Peninsula and its Atlantic surroundings. Guidoboni (1989), in her well researched study of the

[1] Dpto. de Física de la Tierra y Astrofísica, Universidad Complutense, Madrid, Spain. E-mail: audiasva@ucm.es

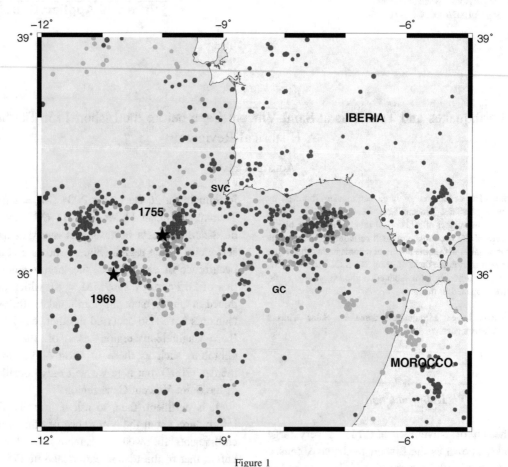

Figure 1
Seismicity of the region of the Saint Vincent Cape for the period 2000–2019 and M > 3.5 taken from Instituto Geográfico Nacional Catalogue. Depth: red < 40 km, green > 40 km. Location of the 1755 and 1969 is given

earthquakes in the Mediterranean region before the year one thousand, does not bring any in the Iberian region. This is explicable, because the Iberian Peninsula was far apart from the cultural centers of Greece and Rome, so it was not of special interest for Greek and Roman historians. The first mention of earthquakes in this region by contemporary authors is only found after the ninth century in some Arabic historians. Regarding those in the Greco-Roman period the first mentions are found in Spanish and Portuguese historians in the sixteenth and seventeenth centuries, such as, Florian de Ocampo (1543), Bernardo de Brito (1597), Juan de Mariana (1601) and Manuel de Faria y Sousa (1628). This information is, then, recorded in the work by Joachim Joseph Moreira de Mendonça (1758), written after the Lisbon earthquake, where the first list of earthquakes in the

Iberian Peninsula is given. This list is used afterwards in the first modern earthquake catalogues for the Iberian Peninsula, specially, by Sánchez Navarro-Neumann (1921), Galbis-Rodriguez (1932), Udías (2017).

1.1. Possible Large Earthquakes and Tsunamis before Year 500 A.D

Historical records can be found about the occurrence of large earthquakes and tsunamis in the region of Saint Vincent Cape in late antiquity. A reference to the oldest of these events can be found in the origin of the legend of the sinking of Atlantis, a large island supposed to have existed to the west of the Strait of Gibraltar (known then as the Pillars of Hercules), the see of a mighty kingdom. The story is included in

Plato's dialogues *Timaeus* and *Critias*, written in the fourth century B.C. *Timaeus* narrates that there were violent earthquakes and flooding and in consequence the island of Atlantis sunk into the sea. *Critias* tells that Atlantis was sunk into the sea by earthquakes. Earthquakes are, then, put at the cause of the sinking of the island. We can, then, assign the origin of this legend to the collective memory of the occurrence in the distant past of some very large earthquakes followed by tsunamis, that flooded the south-western coasts of the Iberian Peninsula. This was afterwards interpreted as islands being sunk into the sea west of the Strait of Gibraltar. The foundation in this region of the city of Cadiz (the old Gadir) by the Phoenicians is estimated to have been around the year one thousand B.C. These earthquakes and tsunamis that gave origin to the legend could have taken place most probably after Cadiz was founded. In *Critias* it is said that the war between the armies of Atlantis before its disappearance into the sea and the Greeks was nine thousand years before its time, but this cannot be used to date the earthquakes. All we can say is that large earthquakes and tsunamis may have occurred in this region in the distant past, giving origin to legends about the disappearance of islands in the sea and the myth of Atlantis.

In the third century B.C., there are some references to earthquakes which caused damage and flooding in Cadiz. They may correspond to large earthquakes followed by tsunamis with origin at the Saint Vincent Cape area. Lacking references in Greek and Roman historians, the first notices are found in the first Spanish historian Ocampo (1543) (Fig. 2), who does not give his sources, and they are repeated by Mariana (1601). The first earthquake is dated by Ocampo (1543) in 241 B.C. saying that "near Cadiz the earth roared and part of the island was flooded" (*cerca de Cádiz bramó la tierra, y anegose parte de la isla*). Mariana (1601) tells in that year of "ordinary earthquakes that it is said to have caused that a part of the island of Cadiz was opened and sunk into the sea" (*ordinarios temblores de tierra, con que una parte de la isla de Cádiz dicen se abrió y se hundió en el mar*). The earthquake appears in the list of Moreira (1758) who takes it from Mariana (1601), but erroneously dates it in 246 B.C. The earthquake figures in the different Spanish catalogues up to the most recent of

Figure 2
Title page of the first edition of Florian de Ocampo, *Crónica General de España* (1543)

Martínez-Solares and Mezcua (2002) dating it, following Moreira, in 246 B.C. Since the oldest reference is that of Ocampo (1543) the correct date must be 241 B.C. Álvarez-Martí-Aguilar (2017) indicates that this is the first reference of an earthquake of the Iberian Peninsula located in Cadiz. The reference to the sinking in the sea of part of the island of Cadiz may indicate that this could have been a very large earthquake followed by a tsunami in the Saint Vincent Cape area.

According to Ocampo (1543) another earthquake in this region took place in 216 B.C. He describes it saying: "The island of Cadiz and all the sea coast of Andalucia suffered large earthquakes which destroyed buildings, killed people, and caused terrible damages and the sea flooded many places" (*La isla de Cádiz y toda la marina frontera de Andalucía padeció grandes terremotos o temblores que derrocaron edificios, mataron gentes y causaron daños terribles y la mar anegó muchos lugares).* Mariana (1601) only says that in the year 218 B.C. there were several earthquakes in Spain. Moreira (1758) follows Mariana (1601) and dates an earthquake in 218 B.C. The catalogue of Martínez-Solares and Mezcua (2002) places an earthquake in SW of Saint Vincent Cape and dates it in 218 B.C., following Moreira (1758). As the first reference is that of Ocampo (1543), the year must be 216 B.C. Here, again, we have reference to earthquakes and flooding in Cadiz, which may correspond to the occurrence of large earthquakes followed by tsunamis in the area of Saint Vincent Cape. Lack of references by more or less contemporary Roman authors could lead to think that Ocampo (1543) mention of the earthquakes of 241 B.C. and 216 B.C. could refer to the same event or events which may have taken place around the middle of the thirteenth century B.C.

The Portuguese historian Manuel de Faria y Sousa (1628) mentions an earthquake in the coast of Portugal the year 60 B.C. that was followed by a tsunami: "The sea surpassing its ordinary limits flooded extended lands" (*El mar excediéndose de sus límites ordinarios cubrió muchas tierras*). Moreira (1758) includes it in his list adding that it happened in Galicia (NW Spain). As a consequence in the next catalogues it figures as taking place in the northern coast of Portugal. However, Galicia is not in the original text of Faria y Sousa (1628) and it could have been anywhere offshore the coast of Portugal. The earthquake is not mentioned by Ocampo (1543) or Mariana (1601) so it is not probable that it is one at Saint Vincent Cape region and felt in Spain.

Moreira de Mendonça (1758) lists an earthquake followed by a tsunami at Saint Vincent Cape in the year 382 A.D., based on the Portuguese historian Brito (1597). Brito bases the occurrence of this earthquake on the testimony of Laymundo, a presumed chaplain of the Spanish Visigoth King Don Rodrigo. But there is no historical evidence of the existence of Laymundo, who is, then, a creation of Brito, therefore the historical base for the earthquake is not valid (Udías 2015). Brito could have had other sources, but they are not given. In spite of this clear lack of historical evidence, the earthquake is listed in Portuguese and Spanish catalogues, for example, Martínez-Solares and Mezcua (2002) giving it as taking place SW of Saint Vincent Cape.

1.2. Large Earthquakes Between Years 500 and 1750

A wealth of information of Arabic authors exists about a large earthquake that took place on the 26th of May 881 (the first earthquake in the Iberian peninsula for which the exact date is known) and probably corresponds to the area of the Saint Vincent Cape. The oldest work where it is mentioned is by Ibn Adhari, a Moroccan author of the thirteenth century which refers to damage in Cordova, the most important Arabian city in South Spain at the time. Mention of the earthquake is also found in the work by Ibn Abi Zar, a Moroccan author of the fourteenth century. The text refers to large damage in the Maghreb and Andalucia, but nothing is said about any phenomenon in the sea. However, the historian José Antonio Conde (1820), based on other Spanish-Arabic texts (no explicit references given), writes about large damage in the south and west coast of Spain and that the sea first retired from the coast and afterwards islands and reefs disappeared in the sea. Mariana (1601) only mentions that in the year 881 there were in the whole of Spain earthquakes with damage and destruction of many buildings. The earthquake is listed by Moreira (1758) and all subsequent catalogues. On this information it could have been a large earthquake in the area of Saint Vincent Cape followed by a tsunami.

The earthquake of 24 August 1356 is a possible one to have occurred in the Saint Vincent Cape area, although there is no information about a tsunami and some catalogues place it in land. Mariana (1601), the first reference, refers to large damage in coastal towns specially Seville and Lisbon. Moreira (1758), quotes Mariana (1601) and refers to damages in Lisbon, Seville, Cordova and other towns in Spain and adds that the earthquake was similar to those of 1531 and 1755. Feijoo (1756) dates it on 23 August, with damage in Lisbon, Algarve and Seville, and refers to the contemporary testimony of Pedro López de Ayala, Major Chancellor of the Kingdom of Castille. Although there is no mention of a tsunami and the earthquake could then have taken place inland, the extend of damage in Lisbon and Seville and the comparison by Moreira de Mendonça (1758) with the earthquakes of 1531 and 1755 may indicate that this could possibly be a large earthquake in the Saint Vincent Cape area, but evidence is not conclusive.

A possible earthquake in the area could also be that of 26 January 1531. It is mentioned, for example, by Mariana (1601) with heavy damage in Lisbon (*tembló horriblemente la tierra en Lisboa*) and the occurrence of a tsunami (*tragose el mar hinchado gran número de navíos*, the swollen sea swallowed up many ships). According to Moreira de Mendonça (1758) it was similar, but even larger than that of 1755, was felt in Africa and it generated a tsunami. However, detailed studies by Justo and Salwa (1998) and Baptista et al. (2014) locate its epicenter at the Bajo Tajo fault and the tsunami restricted to the Lisbon estuary. This interpretation

excludes that the earthquake had its origin at Saint Vincent Cape region but rather somewhere near to the Lisbon area.

2. Discussion

As we have seen, there is historical evidence that very large earthquakes followed by tsunamis have taken place at the Atlantic area west of the Saint Vincent Cape, before the Lisbon earthquake of 1755, causing damages in the western coasts of the Iberian Peninsula and Morocco (Fig. 3). Thus the disastrous 1755 Lisbon earthquake cannot be considered as a unique event, but as part of a series of events, separated by large irregular intervals of time, offshore at the Saint Vincent Cape region or nearby. The largest, of the more recent ones, occurring in 1969 ($M_s = 8$).

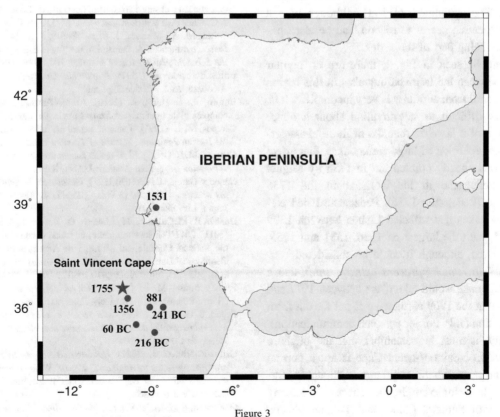

Figure 3
Approximate location of the earthquakes before 1755

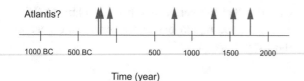

Figure 4
Times of occurrences (red arrows) of the earthquakes before 1755
discussed in the text

The oldest indication of the occurrence of such earthquakes may be the origin of the legend of the sinking of the island of Atlantis, west of the strait of Gibraltar, attributed to earthquakes, as described in Plato's dialogues. Other old earthquakes are those of the third century B.C. (241 and 216), but we do not have contemporary reports, only of later Spanish and Portuguese historians of sixteenth and seventeenth centuries; the historical evidence of their occurrence being then very weak. The earthquake in 881 is well reported by Arabic historians and there is enough information for one in 1356, but this may have taken place inland, since there is no information about a tsunami. The earthquake of 1531, although with a different location nearer to Lisbon, can be also considered as being part of the series.

As can be seen in Fig. 4, there are no regular periods between the large earthquakes in this region and the record from antiquity is very incomplete. It is, then, very difficult to say anything about a future occurrence of a large earthquake of these characteristics. The repetition of large earthquakes, that affect Lisbon, was already pointed out in 1756 by Miguel Tiberio Pedegache in his work about the 1755 earthquake (Pedegache 1756). Pedegache listed eleven earthquakes that affected Lisbon between 1309 and 1755, with the largest on 1356, 1531 and 1755, and predicted, although it could be considered "extravagant" (*a muitos parecerá extravagante*), that a large earthquake would take place between 1977 and 1985, so that the 1969 earthquake is not far out from his prediction (J. P. Poirier personal communication).

Finally, it must be remember that the offshore region west of Saint Vincent Cape is not a normal subduction zone, where very large earthquakes usually take place, for example, the recent ones already mentioned of Sumatra (2004) and Tohoku (2011). The origin of the occurrence of very large

earthquakes, as the one of 1755, in this region is still a debated question.

Acknowledgements

This work was partially supported by the Spanish Ministerio de Economía, Industria y Competitividad, project CGL2017-86097-R.

Publisher's Note   Springer Nature remains neutral with regard to jurisdictional claims in published maps and institutional affiliations.

REFERENCES

Alonso, C., Gracia, F. J., Rodríguez-Polo, S., & Martín Puertas, C. (2015). El registro de eventos energéticos marinos en la bahía de Cádiz durante épocas históricas. *Cuaternario y Geomorfología, 29*(1–2), 95–117.

Álvarez-Martí-Aguilar, M. (2017). La tradición historiográfica sobre catástrofes naturales en la Península Ibérica durante la Antigüedad y el supuesto tsunami del Golfo de Cádiz de 218–209 A.C. *Dialogues d'histoire ancienne, 43*, 117–145.

Baptista, M. A., Miranda, J. M., & Batlló, J. (2014). The 1531 Lisbon earthquake: A Tsunami in the Tagus Estuary. *Bulletin of the Seismological Society of America, 104*, 2149–2161.

Brito, Bernardo de (1597). *Monarchia Lusytana*. Mosteiro de Alcobaça: Alexander de Siquiera.

Buforn, E., & Udías, A. (2010). Azores-Tunisia, a tectonically complex plate boundary. *Advances in Geophysics, 52*, 139–182.

Campos, M. L. (1991). Tsunami hazard on the Spanish Coasts of the Iberian Peninsula. *Science of Tsunami Hazards, 9*, 83–90.

Campos, M. L. (1992). *El riesgo de tsunamis en España: análisis y valoración geográfica*. Madrid: I.G.N./MOPT.

Conde y García, J. A. (1820,1821), *Historia de la dominación de los árabes en España*. (3 vols.) (Edición de 1844), Barcelona: Imp. y Lib. Española.

Dunbar, P., McCullough, H., Mungov, G., Varner, J., & Stroker, K. (2011). 2011 Tohoku earthquake and tsunami data available from the National Oceanic and Atmospheric Administration/National Geophysical Data Center. *Geomatics, Natural Hazards and Risk, 2*(4), 305–323. https://doi.org/10.1080/19475705.2011.632443.

Faria y Sousa, M. de (1628). *Historia del Reino de Portugal* (Nueva Edición 1730). Brussels: Francisco Foppens.

Feijoo, B. G. (1756). *Nuevo sisthema sobre la causa physica de los terremotos explicado por los phenomenos eléctricos*. Lisbon: Joseph da Costa Coimbra.

Galbis-Rodriguez, J. (1932). *Catálogo sísmico de la zona comprendida entre los meridianos 5° E y 20° W de Greenwich y los paralelos 45° y 25 N*. Tomo I. Madrid: Instituto Geográfico Catastral y Estadístico.

Guidoboni, E. (Ed.). (1989). *I terremoto prima del mille in Italia e nell'area Mediterranea*. Bologna: Storia-Geofisica-Ambiente.

Ibn Adhar (1963). *Al-Bayano'l-Mogrib (Spanish translation by A. Huici Miranda, in Al-Bayán al-Mugrib. Nuevos fragmentos almorávides y almohades)*. Valencia: Anubar.

Ibn Abi Zar (1964). *Rawd al-Qirtas (Spanish translation by A. Huici Miranda, Rawd el-Qirtas)*. Valencia: Anubar.

Justo, J. L., & Salwa, C. (1998). The 1531 Lisbon earthquake. *Bulletin of the Seismological Society of America, 88,* 319–328.

Mariana, Juan de (1601). *Historia General de España*. Biblioteca de Autores Españoles vols. 30 y 31. Madrid: Rivadeneira (1854).

Martínez-Solares, J. M., & Mezcua, J. (2002). *Catálogo sísmico de la península Ibérica (880 A.C.–1900). Monografía 18*. Madrid: Instituto Geográfico Nacional.

Moreira de Mendonça, J. J. (1758), *Historia Universal dos terremotos*. Lisboa: A.V. da Silva.

Ocampo, Florian de (1543), *Crónica General de España*. (Edition used of 1791, Madrid: Benito Cano).

Pedegache, M. T. (1756). *Nova e fiel relaçaõ de terremoto que experimentou Lisboa e todo Portugal no 1 de November de 1755 com algunas observaçoens curiosas e a explicaõ das suas causas*. Lisbon: Manuel Soares.

Platón. (1992). *Diálogos, VI, Filebo, Timeo, Critias* (Traducciones, introducciones y notas por Mª Ángeles Durán y Francisco Lisi). Madrid: Gredos.

Sánchez Navarro-Neumann, J. M. (1921). *Bosquejo sísmico de la península Ibérica* (Lista de los terremotos destructores sentidos en la península Ibérica). Granada: La Estación Sismológica y Observatorio Astronómico de Cartuja.

Udías, A. (2015). Historical earthquakes (before 1755) of the Iberian peninsula in early catalogs. (Electronic Supplement: Critical revision of earthquakes in the Iberian peninsula before year 1000). *Seismological Research Letters, 86,* 999–1005.

Udías, A. (2017). Terremotos de la Península Ibérica antes de 1900 en los catálogos sísmicos. *Física de la Tierra, 29,* 11–27.

(Received June 24, 2019, revised September 6, 2019, accepted September 9, 2019, Published online September 16, 2019)

Pure Appl. Geophys. 177 (2020), 1747–1759
© 2019 Springer Nature Switzerland AG
https://doi.org/10.1007/s00024-019-02340-y

▌Pure and Applied Geophysics

Seismic Information at the Spanish Geophysical Data National Archive. Example: the Earthquake of February 28, 1969

MARINA LÓPEZ MUGA,[1] IRENE BENAYAS,[1] and JOSE MANUEL TORDESILLAS[1]

Abstract—The Spanish Geophysical Data National Archive (Toledo Geophysical Observatory) is a centre created by *Instituto Geográfico Nacional* to collect all the geophysical documentation produced in all of the observatories IGN has had in operation throughout its history. All this information is transferred to the Geophysical Data National Archive where it is reviewed, classified and catalogued in a database, to be eventually filed in the records repository of the Archive, with the appropriate conditions for their future preservation. Its contents are digitized as backup and to meet the data requests received in this Archive. For the study of earthquakes throughout the twentieth century, this Archive has a large volume of information. The documents most consulted by researchers are the seismic records from the IGN Geophysical Observatories, but also the Archive houses an important collection of complementary information.

Key words: Seismology, historical seismic data, seismic archive.

1. Introduction

The Spanish Geophysical Data National Archive (SGDNA) is a centre created in 2007 by *Instituto Geográfico Nacional* of Spain (IGN), located within Toledo Geophysical Observatory facilities.

Its purpose is to centralise all the geophysical documentation produced in every observatory IGN has had in operation for most of the twentieth century (Table 1), which corresponds mainly with Seismological, Geomagnetic and Geoelectric data, although it is also reinforced with complementary documentation like Meteorological data, etc.

These data are registered upon analogical formats—mainly paper of different types—although there is also a surplus of microfilms, photographs, slides, etc.

Until the creation of SGDNA, all data stemmed from the different observatories were stored in their facilities, despite lacking an adequate environment suitable for its proper maintenance and future preservation.

After its creation, the Geophysical Data National Archive received all the documentation which was thoroughly labelled and organised for proper storage in an adequate environment for safe preservation.

The SGDNA has a 103 m² conditioned-records repository with compact file cabinets, 1240 m of total storage capacity (Fig. 1). Said repository is located in the basement, where environmental conditions are kept at low temperature variation and regulated humidity, thus protecting data for a proper preservation.

Documents are stored in special file boxes with an inner neutral pH covering to avoid oxidation and a lining index less than 1% to prevent potential damage by cellulose, and made with appropriate formats for each type of document.

This paper describes the seismic information that has been stored in the SGDNA, which has been assigned a value for the purposes of future research by the scientific community.

2. Seismic Data in the Archive

Seismic information at the Geophysical Data National Archive is quite abundant and diverse. It ranges from the beginnings of IGN observatories in

[1] Archivo Nacional de Datos Geofísicos, Instituto Geográfico Nacional, Observatorio Geofísico de Toledo, Avenida Adolfo Suárez, km. 4, Toledo 45005, Spain. E-mail: mlmuga@fomento.es; ibenayas@gmail.com; jmtordesillas@fomento.es

Table 1

List of IGN observatories

Observatory	Data recollection period	Disciplines
Alicante Seismological Observatory (ALI)	1914–1985	Seismology
Almería Geophysical Observatory (ALM)	1911–1990	Seismology and Geomagnetism
Logroño Geophysical Observatory (LGR)	1962–1998	Seismology and Geomagnetism
Málaga Seismological Observatory (MAL)	1913–1985	Seismology
Santa Cruz de Tenerife Geophysical Observatory (TEN)	1963–1999	Seismology and Geomagnetism
Toledo Geophysical Observatory (TOL)	1909–2000	Seismology, Geomagnetism and Geoelectricity
Santiago de Compostela Observatory (STS)	1976–1984	Seismology
San Pablo de los Montes Geophysical Observatory (SPT)	1992–2006	Seismology and Geomagnetism
Sonseca Seismic Array	1962–1981	Seismology

Figure 1

Records repository of the SGDNA, showing records catalogued and permanently stored in appropriate environmental conditions for their preservation

1909 to the end of analogue data by the late twentieth century when the remaining observatories were updated to digital data.

The seismic information has been grouped in several areas that will be described as follows.

2.1. Historical Seismograms

The first group of seismic information stored in the SGDNA is composed of historical seismograms recorded at seismic observatories (Table 2). This is the largest collection of the archive and perhaps the most important for the scientific community.

From the archival perspective, this documentation is very special due to the variety of formats used throughout history to record seismograms. In the

Archive recorded seismograms can be found upon the following formats:

- Smoked paper
- Photographic paper
- Ink paper
- Thermal paper
- Audio-visual records (film record, microfilm and photographic record).

Another peculiarity of such data is how diverse the formats of the documents are. The size ranges from 12×7 cm—as in WWSSN microfilms—to 240×15 cm—for Vicentini's smoked paper seismograms—besides the flat films records of Sonseca Seismic Array.

Because of these characteristics, there was the need to design different types of archives boxes and storage devices that would fulfill the requirements for an adequate conservation of seismograms.

The most important part of the collection is composed by the IGN seismograms recorded in their observatories (close to a million of documents) and data recorded in Sonseca Seismic Array (13375 flat films).

Seismograms received at the SGDNA are duly identified and sorted. If some of them are slightly deteriorated they undergo a recovery process. The cataloguing criterion applied to the sorting of seismograms is respecting and following their original sequence: by seismic station, by instruments and by chronological order. Once the seismograms are catalogued and stored in the records repository of the Archive, the information about the location of each

Table 2

Periods for which seismic records of IGN observatories are available

Instrument	Station							
	ALI	ALM	MAL	LGR	STS	TEN	TOL	SPT
Vicentini	1914–1923	1911–1937	1915–1935	–	–	–	1909–1923	–
Agamennone	–	–	–	–	–	–	1909–1924	–
Bosch-Omori	1914–1923	1911–1928	1915–1924	–	–	–	1909–1922	–
Milne	–	–	–	–	–	–	1909–1922	–
Rebeur-Ehlert	–	–	–	–	–	–	1909–1914	–
Wiechert Astático	–	–	–	–	–	–	1910–1993	–
Wierchert Toledo	–	–	–	–	–	–	1931–1981	–
Wiechert Vertical	1921–1976	–	1924–1941	–	–	–	1924–1987	–
Wiechert-Guillamon			1941–1960	–	–	–	–	–
Mainka Horizontal	1924–1975	1928–1985	1924–1953	–	–	–	–	–
Mainka Vertical	–	1926–1972	–	–	–	–	–	–
Victoria	–	1953–1959	1940–1960	–	–	–	–	–
Galbis	1935–1958	–	–	–	–	–	–	–
Hiller-Stuttgart	1962–1985	1959–1990	1960–1965	1962–1996	–	1962–1983	–	–
Sprengnether	–	–	–	–	–	–	1958–1987	–
WWSSN	–	–	1962–1997	–	–	–	–	–
Kinemetrics	–	–	–	–	–	1988–2013	–	1992–2006
Lennartz	–	–	–	–	1976–1984	1976–2003	–	–

seismogram is loaded in the Archive's database by storage units (box identifier) for easy identification.

It is necessary to highlight the seismograms digitization project that is being carried out, by means of which the seismograms with earthquakes signals of magnitude superior to 3.5 are being digitized in the case of nearby earthquakes and of magnitude superior to 6.0 in the case of the distant ones.

The digitization process is being carried out with A0 size overhead scanners. Each image obtained covers a complete seismogram in TIFF format at 300 optical dpi resolution, with increasing resolution up to 1200 dpi for the most interesting fragments of the seismograms.

Besides the original seismograms of IGN, the SGDNA also has an interesting collection of seismograms copies courtesy of foreign observatories, usually in microfilm format, slideshows, or paper copies, and they highlight particular events more than an actual series.

2.2. Seismograms Analyses

The following data group concerns seismograms analyses made at IGN observatories when recording an earthquake. They are documents that collect relevant data like original readings of seismic waves, arrival times and their amplitudes.

The original data was the background for seismic bulletins, and is currently very useful. In some instances, bulletins covering a specific earthquake are lost, or in some cases possible transcriptions errors arise when anomalous values are present in bulletins, therefore such seismogram analyses could provide more information about that earthquake recorded in particular.

Because they are manuscripts originally produced at Observatories, these are considered very valuable, and for such reason are being digitized to avoid further deterioration due to handling and to have a copy in case of a potential loss.

The SGDNA houses a large collection of said data, official property of the IGN Observatories.

Figure 2
Seismic bulletins collection of the SGDNA, currently in process of cataloguing and digitization. After that process, they are stored in special boxes in order to protect and preserve them

2.3. Sesimic Bulletins

Monthly seismic bulletins comprise another set of information on the earthquakes recorded and reported in that period. Historically, Observatories shared and circulated among themselves the bulletins each one produced.

These bulletins contain information about seismic phases, magnitude, intensity, location and victims of the earthquake recorded.

SGDNA actually works on a compilation of all seismic bulletins both produced at IGN Observatories and the ones received from other Observatories, Spanish or foreign institutions. The large volume the Archive comes from a variety of worldwide Centres, which is currently being catalogued, digitized and stored in the records repository of the SGDNA (Fig. 2).

Clearly, the vast majority of documents of this type comprise seismic bulletins of IGN Observatories, both provisional and definitive bulletins, which are preserved in many cases. SGDNA has two distinctive groups of IGN bulletins: those from Observatories where nearby and distant earthquakes were indicated, and the ones from IGN headquarters

only indicating nearby earthquakes. These last bulletins are fully digitized and posted on the IGN website. Currently, bulletins of different Observatories are being digitized for their eventual post on said website. All bulletins digitized are being catalogued and filed in the Archive.

2.4. Marcroseismic Questionnaires

The Marcroseismic Questionnaires are documents describing how an earthquake has been experienced as reported by people in the affected area and sent to Geophysical Observatories or to IGN headquarters. All available questionnaires at SGDNA come from Spanish localities.

This kind of information is crucial in order to establish the earthquake seismic intensity in every place and precedes the elaboration of Isoseismal Maps.

Many of these questionnaires are classified by earthquake and in chronological order, then, for preservation purposes, stored in special archive boxes (Fig. 3) and filed in stable polyester folders free of plasticizers, residual solvents and active sulphides

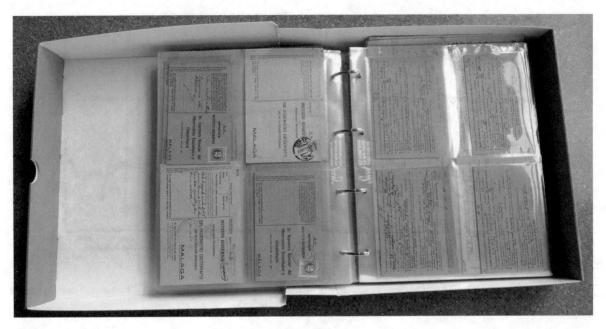

Figure 3
Macroseismic questionnaires catalogued and permanently stored. Each questionnaire is filed in a special polyester folder in order to be preserved, and all the folders are stored in special archive boxes

components, with neutral and chemically inert pH, that has passed the silver tarnish test and the solvent activity test.

The SGDNA also has a directory of macroseismic informants from each city or town. So, in those cases where a questionnaire does not name the location in question, we can provide it by deducting the sender's profile and looking directly at the source in our own directory.

So far, 825 seismic questionnaires have been catalogued by the SGDNA, the oldest one from the Santa Pola (Alicante) earthquake in July 1st, 1920.

2.5. Isoseismal Maps

The next group of documentation the SGDNA stores are the Isoseismal Maps of main earthquakes. This maps show graphically the intensity values from any given location of every specific earthquake.

This is not standardised information given the assorted variety of papers in several sizes and scales. Nevertheless, they can be useful for different studies when other macroseismic information about an earthquake is lost or missing.

The SGDNA have a substantial quantity of isoseismal maps that has been widely catalogued and preserved in map cases following the proper archival regulations for this kind of documents. The oldest Isoseismal Map in the SGDNA is that of the earthquake (Fig. 4) of May 25, 1901, Motril (Granada) being the epicentre.

There are future plans to digitize this whole collection of isoseismal maps and publish it on the IGN website.

2.6. Historic Photographs

Another important collection the SGDNA owns is that of historic photographs (since 1884). They come in different formats: glass, films, paper, slideshows, etc., which include views of Observatories and their instrumentation, and damages produced by earthquakes.

Photographs undergo a process that reveals their content. If instruments are shown the next step is identifying the Observatory where these were installed. In the case of damages produced in

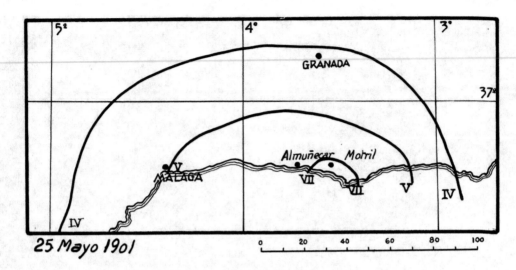

Figure 4
Isoseismal map of the May 25, 1901 earthquake with epicentre in Motril, Granada. It corresponds to the oldest isoseismal map that can be found currently in the SGDNA

Figure 5
Glass plates installed in especial boxes. In order to preserve them, each one is stored in a special cover made to measure in the SGDNA. Then they are placed in special boxes to ensure their preservation

buildings or cities depicted in photographs, then it's necessary to find out which earthquake is the culprit.

After proper identification, the photograph is digitized and filed accordingly. It's worth mentioning that given the various formats used, each one is filed separately in different archives boxes or albums according to the preservation treatment it will undergo (Fig. 5).

Up to this time, a total of 3667 photographs has been catalogued and uploaded to the SGDNA database. The distribution according to their different formats is:

Photographs on paper → 629
Photographs on glass → 107
Photographs on nitrate film and cellulose acetate film → 107
Slideshows → 2824.

SGDNA digitized all these filed documents, including collections loaned by other institutions or private individuals, all with seismic-related content.

2.7. Press News

The SGDNA also has an important set of news clips from newspapers covering important earthquakes.

IGN and IGN Observatories in particular, keep an important collection of news clips (since 1913) and already part of the Archive. All the ones SGDNA houses have been digitized and kept in special boxes.

Besides these filed news clip, the Archive attempts to increase the collection with digital copies from Newspapers Archives. The SGDNA is joining efforts with the main newspaper archives in order to gather as many news clips as possible featuring coverage of the main earthquakes occurred in our area.

2.8. External Archives Data

For the making of digital data repository that envelopes all knowledge in relation to any significant earthquake occurred in Spain and nearby areas, the SGDNA is consulting with External Archives on further information about specific earthquakes. In many cases these Archives are being very cooperative and share whatever information they possess.

Overall, three groups of Archives can be considered for filing purposes: Administrative, Foreign and Scientific.

Among the Administrative Archives there are three categories:

- Municipal Archives whose cities were affected by an earthquake.
- Historical Archives

- General Archive of the Administration (AGA), gathering information from files on monument restoration or Memories of Civil Governments.

The SGDNA also has obtained information of foreign archives from neighbouring countries, and scientific archives that may include seismic-related content.

3. The SGDNA Database

In order to manage all this seismic-related data, the SGDNA has designed its own database with files featuring specific earthquakes for future reference.

Given the amount of data SGDNA deals with, this is a very ambitious project for an understaffed institution with extra work load, but nevertheless able to manage and achieve its goal in locating as much information as possible in order to meet the researchers needs.

4. The Saint Vicent Cape Earthquake (February 28, 1969): A Case of Study

Below is a detailed description of the data available in the Geophysical Data National Archive about the Saint Vincent Cape earthquake that took place on February 28, 1969 (Table 3), widely felt in Morocco, Portugal and Spain. It has been the last large earthquake in the Iberian Peninsula and Morocco (Martínez Solares 2011).

4.1. Seismograms

The main documentation on this earthquake SGDNA has corresponds to the seismograms of such event. Table 4 shows a list of all the seismograms that are located up to this moment. Due to the earthquake magnitude, most of the seismic stations of

Table 3

Parameters of February 28, 1969 earthquake

Date	Time (UTC)	Latitude	Longitude	Depth	Magnitude	Intensity	Location
28/02/1969	02:40:32	35.9850	− 10.8133	20	7.8 (Mw)	VIII	SW SAINT VINCENT CAPE

Table 4

Seismograms of February 28, 1969 earthquake, at SGDNA

Observatory	Seismograph	Components	Support material
Toledo Geophysical Observatory (TOL)	Spregnether (belonging to WWSSN)	N–S, E–W, Z	Photographic paper
	Benioff (belonging to WWSSN)	N–S, E–W, Z	Photographic paper
	Wiechert Astatic	N–S, E–W	Smoked paper
	Wiechert Vertical	Z	Smoked paper
	Wiechert Toledo	E–W	Smoked paper
Alicante Seismological Observatory (ALI)	Mainka	N–S, E–W	Smoked paper
	Hiller-Stuttgart	N–S, E–W, Z	Photographic paper
Almería Geophysical Observatory (ALM)	Mainka Vertical	Z	Smoked paper
Sonseca Array (ZEBU)	Short-period stations		Flat film
	Long-period stations		Flat film
Umea, Sweden (UME)	Spregnether (belonging to WWSSN)	N–S, E–W, Z	Microfiche
	Benioff (belonging to WWSSN)	N–S, E–W, Z	Microfiche
Port Moresby, New Guinea (PMG)	Spregnether (belonging to WWSSN)	N–S, E–W, Z	Microfiche
	Benioff (belonging to WWSSN)	N–S, E–W, Z	Microfiche
Kevo, Finland (KEV)	Benioff (belonging to WWSSN)	N–S, E–W, Z	Microfiche
Kajaani, Finland (KJN)	Benioff (belonging to WWSSN)	N–S, E–W, Z	Microfiche
Nurmijarvi, Finland (NUR)	Benioff (belonging to WWSSN)	N–S, E–W, Z	Microfiche
Santo Domingo (Ciudad Trujillo) Dominican Republic (SDD)	Benioff (belonging to WWSSN)	Z	Microfiche
Nana, Peru (NNA)	Spregnether (belonging to WWSSN)	N–S, E–W, Z	Microfiche
	Benioff (belonging to WWSSN)	N–S, E–W, Z	Microfiche

the Iberian Peninsula were saturated, but these seismograms are still being useful for many studies because of the information they contain such as arrivals of the first phases, records of aftershocks, etc.

4.2. Seismograms Analysis and Telegrams

Another set of very valuable documentation about the earthquake being studied is formed by the seismograms analysis (Fig. 6) that were made in a preliminary stage of the seismic bands, and the telegrams that were sent between different observatories sharing the information registered at first.

Table 5 shows a list of the documents of this type from SGDNA, where the original calculations related to the seismograms of Alicante, Tenerife, Málaga and Moca are kept, providing insight information as the seismic bands of these Observatories are misplaced.

4.3. Seismic Bulletins

Due to its great magnitude, this event was recorded by a large number of seismic stations and therefore phases of their recordings can be found in multiple bulletins from around the world.

The list of seismic bulletins available in the SGDNA where readings of this event appear is too long for this article to deal with. For that reason, only some bulletins from institutions closer to the epicentre will be highlighted (Table 6) as they provide information on how the earthquake was felt by different populations, and they even include isoseismal maps.

4.4. Macroseismic Questionnaires

The earthquake in question was widely felt in Spain, Portugal and Morocco so many macroseismic questionnaires were received at the IGN and its observatories. They were used for large number of studies, but eventually misplaced as time went by.

Today, SGDNA has been able to retrieve only 9 macroseismic questionnaires for this earthquake, courtesy of the Málaga Observatory (Fig. 7), and they come from the following populations:

- Cartaya (Huelva)
- Cortegana (Huelva)
- Puebla de Guzmán (Huelva)
- Morón de la Frontera (Seville)

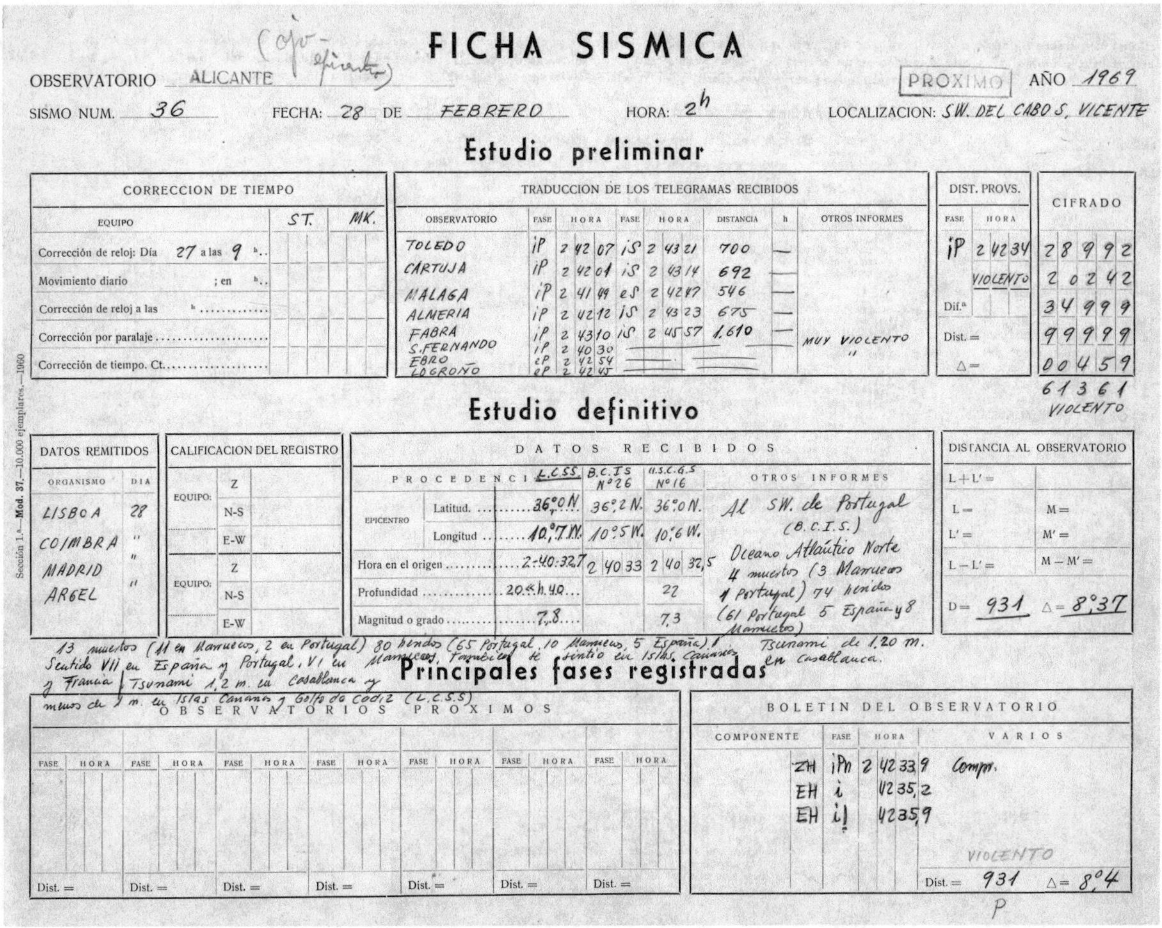

FICHA SISMICA

OBSERVATORIO ALICANTE (Ojo - efectivo) PROXIMO AÑO 1969

SISMO NUM. 36 FECHA: 28 DE FEBRERO HORA: 2h LOCALIZACION: SW. DEL CABO S. VICENTE

Estudio preliminar

CORRECCION DE TIEMPO

EQUIPO	ST.	MK.
Corrección de reloj: Día 27 a las 9 h..		
Movimiento diario ; en h..		
Corrección de reloj a las h		
Corrección por paralaje		
Corrección de tiempo. Ct....................		

TRADUCCION DE LOS TELEGRAMAS RECIBIDOS

OBSERVATORIO	FASE	HORA	FASE	HORA	DISTANCIA	h	OTROS INFORMES
TOLEDO	iP	2 42 07	iS	2 43 21	700	—	
CARTUJA	iP	2 42 01	iS	2 43 14	692	—	
MALAGA	iP	2 41 49	eS	2 42 47	546	—	
ALMERIA	iP	2 42 12	iS	2 43 23	675	—	
FABRA	iP	2 43 10	iS	2 45 57	1.610	—	MUY VIOLENTO
S.FERNANDO	iP	2 40 30					"
EBRO	eP	2 42 54					
LOGROÑO	eP	2 42 45					

DIST. PROVS.

FASE	HORA
iP	2 42 34
	VIOLENTO

Dif.a

Dist. =

Δ =

CIFRADO

2 8 9 9 2
2 0 2 4 2
3 4 9 9 9
9 9 9 9 9
0 0 4 5 9
6 1 3 6 1
VIOLENTO

Estudio definitivo

DATOS REMITIDOS

ORGANISMO	DIA
LISBOA	28
COIMBRA	"
MADRID	"
ARGEL	"

CALIFICACION DEL REGISTRO

EQUIPO:	Z
	N-S
	E-W
EQUIPO:	Z
	N-S
	E-W

DATOS RECIBIDOS

PROCEDENCIA	L.C.S.S.	B.C.I.S Nº26	U.S.C.G.S Nº16
EPICENTRO Latitud	36°0 N.	36°2 N.	36°0 N.
EPICENTRO Longitud	10°7 W.	10°5 W.	10°6 W.
Hora en el origen	2-40:32,7	2 40 33	2 40 32,5
Profundidad	20 km. h.40		22
Magnitud o grado	7,8		7,3

OTROS INFORMES: Al SW. de Portugal (B.C.I.S.) Oceano Atlántico Norte 4 muertos (3 Marruecos 1 Portugal) 74 heridos (61 Portugal 5 España y 8 Marruecos)

DISTANCIA AL OBSERVATORIO

L + L' =

L = M =

L' = M' =

L − L' = M − M' =

D = 931 Δ = 8°37

13 muertos (11 en Marruecos, 2 en Portugal) 80 heridos (65 Portugal, 10 Marruecos, 5 España). Tsunami de 1,20 m. Sentido VII en España y Portugal, VI en Marruecos, también se sintió en Islas Canarias y Francia. Tsunami 1,2 m. en Casablanca y menos de 1 m. en Islas Canarias y Golfo de Cadiz (L.C.S.S.) en Casablanca.

Principales fases registradas

OBSERVATORIOS PROXIMOS

FASE	HORA	FASE	HORA	FASE	HORA	FASE	HORA	FASE	HORA	FASE	HORA	FASE	HORA
Dist. =		Dist. =		Dist. =		Dist. =		Dist. =		Dist. =		Dist. =	

BOLETIN DEL OBSERVATORIO

COMPONENTE	FASE	HORA	VARIOS
ZH	iPn	2 42 33,9	Compr.
EH	i	42 35,2	
EH	i	42 35,9	
			VIOLENTO
Dist. = 931			Δ = 8°4

P

Sección 1.ª—Mod. 37.—10.000 ejemplares.—1960

Figure 6

Seismograms analysis of February 28, 1969 earthquake. It was made at Alicante Observatory and shows information received by telegram from other Observatories of the Iberian Peninsula, and their own measurements and calculations

Table 5

Seismogram analysis preserved at the SGDNA

Observatory	Type of files
Alicante Seismological Observatory	Seismograms analysis
Málaga Seismological Observatory	Preliminary calculation
Santa Cruz de Tenerife Geophysical Observatory	Seismograms analysis
Moca Geophysical Observatory (Equatorial Guinea)	Preliminary calculations

Table 6

Seismic bulletins of February 28, 1969 earthquake

Institution	Country
Faculté des Sciences. Institut scientifique Cherifien	Rabat, Morocco
Institut de Meteorologie et de Physique Du Globe d'Algerie	Algeria
Serviço Meteorológico Nacional de Portugal	Lisbon, Portugal
Instituto Nazionale di Geofisica de Italia	Roma, Italy
Real Observatorio de la Armada (ROA)	Cádiz, Spain
Observatori del Ebro	Tortosa, Spain

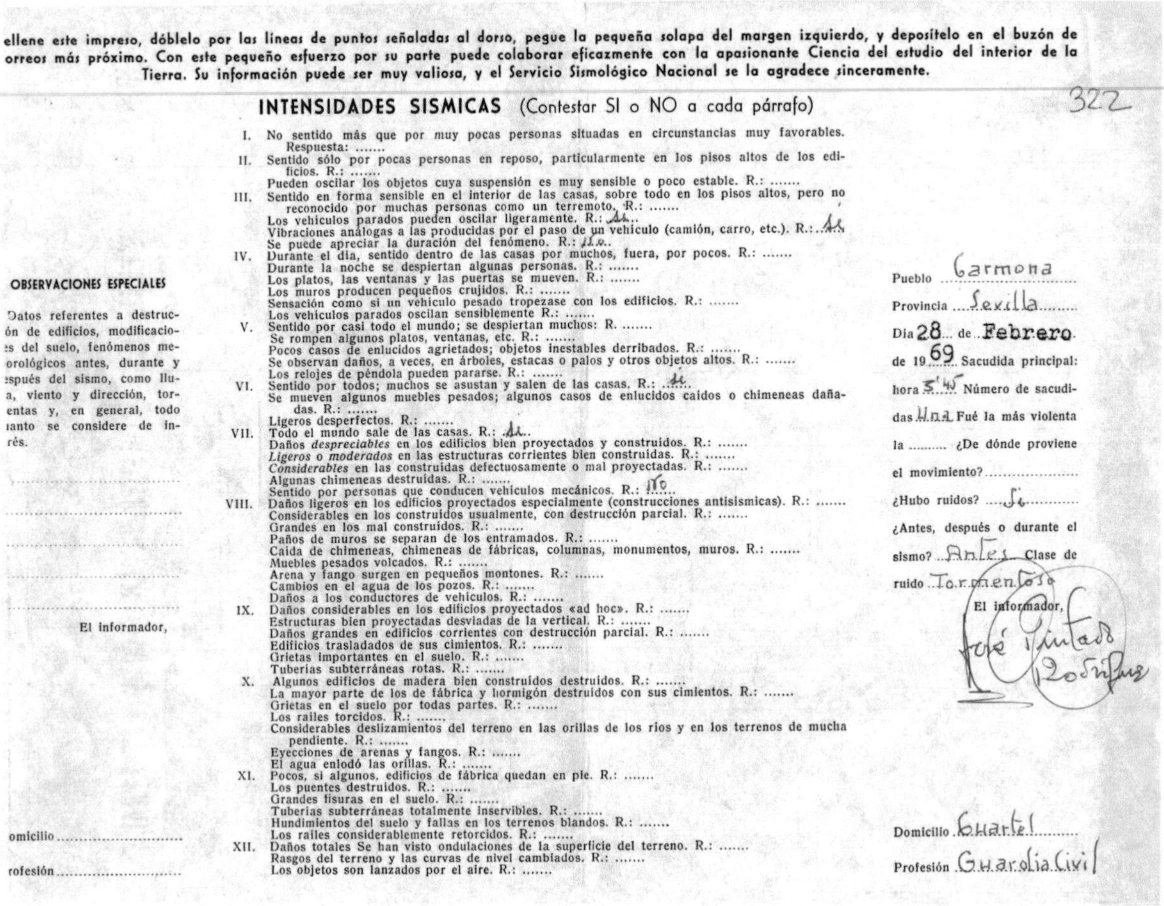

ellene este impreso, dóblelo por las líneas de puntos señaladas al dorso, pegue la pequeña solapa del margen izquierdo, y depositelo en el buzón de orreos más próximo. Con este pequeño esfuerzo por su parte puede colaborar eficazmente con la apasionante Ciencia del estudio del interior de la Tierra. Su información puede ser muy valiosa, y el Servicio Sismológico Nacional se la agradece sinceramente.

322

INTENSIDADES SISMICAS (Contestar SI o NO a cada párrafo)

I. No sentido más que por muy pocas personas situadas en circunstancias muy favorables. Respuesta: ....

II. Sentido sólo por pocas personas en reposo, particularmente en los pisos altos de los edificios. R.: ....
Pueden oscilar los objetos cuya suspensión es muy sensible o poco estable. R.: ....

III. Sentido en forma sensible en el interior de las casas, sobre todo en los pisos altos, pero no reconocido por muchas personas como un terremoto. R.: ....
Los vehiculos parados pueden oscilar ligeramente. R.: si
Vibraciones análogas a las producidas por el paso de un vehiculo (camión, carro, etc.). R.: si
Se puede apreciar la duración del fenómeno. R.: no

IV. Durante el dia, sentido dentro de las casas por muchos, fuera, por pocos. R.: ....
Durante la noche se despiertan algunas personas. R.: ....
Los platos, las ventanas y las puertas se mueven. R.: ....
Los muros producen pequeños crujidos. R.: ....
Sensación como si un vehiculo pesado tropezase con los edificios. R.: ....
Los vehiculos parados oscilan sensiblemente R.: ....

V. Sentido por casi todo el mundo; se despiertan muchos: R. ....
Se rompen algunos platos, ventanas, etc. R.: ....
Pocos casos de enlucidos agrietados; objetos inestables derribados. R.: ....
Se observan daños, a veces, en árboles, estacas o palos y otros objetos altos. R.: ....
Los relojes de péndola pueden pararse. R.: ....

VI. Sentido por todos; muchos se asustan y salen de las casas. R.: si
Se mueven algunos muebles pesados; algunos casos de enlucidos caidos o chimeneas dañadas. R.: ....
Ligeros desperfectos. R.: ....

VII. Todo el mundo sale de las casas. R.: si
Daños *despreciables* en los edificios bien proyectados y construidos. R.: ....
Ligeros o *moderados* en las estructuras corrientes bien construidas. R.: ....
Considerables en las construidas defectuosamente o mal proyectadas. R.: ....
Algunas chimeneas destruidas. R.: ....
Sentido por personas que conducen vehiculos mecánicos. R.: no

VIII. Daños ligeros en los edificios proyectados especialmente (construcciones antisismicas). R.: ....
Considerables en los construidos usualmente, con destrucción parcial. R.: ....
Grandes en los mal construidos. R.: ....
Paños de muros se separan de los entramados. R.: ....
Caida de chimeneas, chimeneas de fábricas, columnas, monumentos, muros. R.: ....
Muebles pesados volcados. R.: ....
Arena y fango surgen en pequeños montones. R.: ....
Cambios en el agua de los pozos. R.: ....
Daños a los conductores de vehiculos. R.: ....

IX. Daños considerables en los edificios proyectados «ad hoc». R.: ....
Estructuras bien proyectadas desviadas de la vertical. R.: ....
Daños grandes en edificios corrientes con destrucción parcial. R.: ....
Edificios trasladados de sus cimientos. R.: ....
Grietas importantes en el suelo. R.: ....
Tuberias subterráneas rotas. R.: ....

X. Algunos edificios de madera bien construidos destruidos. R.: ....
La mayor parte de los de fábrica y hormigón destruidos con sus cimientos. R.: ....
Grietas en el suelo por todas partes. R.: ....
Los railes torcidos. R.: ....
Considerables deslizamientos del terreno en las orillas de los rios y en los terrenos de mucha pendiente. R.: ....
Eyecciones de arenas y fangos. R.: ....
El agua enlodó las orillas. R.: ....

XI. Pocos, si algunos, edificios de fábrica quedan en pie. R.: ....
Los puentes destruidos. R.: ....
Grandes fisuras en el suelo. R.: ....
Tuberias subterráneas totalmente inservibles. R.: ....
Hundimientos del suelo y fallas en los terrenos blandos. R.: ....
Los railes considerablemente retorcidos. R.: ....

XII. Daños totales Se han visto ondulaciones de la superficie del terreno. R.: ....
Rasgos del terreno y las curvas de nivel cambiados. R.: ....
Los objetos son lanzados por el aire. R.: ....

OBSERVACIONES ESPECIALES

Datos referentes a destrucón de edificios, modificacioes del suelo, fenómenos meorológicos antes, durante y espués del sismo, como llua, viento y dirección, torentas y, en general, todo uanto se considere de inrés.

El informador,

omicilio ....

rofesión ....

Pueblo Carmona

Provincia Sevilla

Dia 28 de Febrero de 1969 Sacudida principal: hora 5'45 Número de sacudidas Una Fué la más violenta la .... ¿De dónde proviene el movimiento? ....

¿Hubo ruidos? Si

¿Antes, después o durante el sismo? Antes Clase de ruido Tormentoso

El informador,

José Pintado Rodríguez

Domicilio Cuartel

Profesión Guardia Civil

Figure 7
Macroseismic questionnaire from Carmona, Seville, corresponding to the February 28, 1969 earthquake. It gives information on how the earthquake felt in this population by people

- Carmona (Seville)
- Marchena (Seville)
- Valverde del Camino (Huelva)
- Cabeza la Vaca (Badajoz)
- Bélmez (Córdoba).

4.5. Photographs

Despite SGDNA extensive photographic material, the institution lacks the original photographs of this earthquake in particular, but lately was able to locate a set of photographs—taken by different newspaper agencies—that were kept in other Administration Archives.

That is the case of four photographs (set 6359) provided by the Municipal Archive of Isla Cristina, Huelva, showing the damaged church of *Nuestra Señora de los Dolores* (Fig. 8).

Eight photos of damages suffered in Seville as a result of the earthquake have also been recovered. These have been obtained from the General Archive of the Administration (F/01456) and belong to the CIFRA agency (Fig. 9).

4.6. Administrative Documents

For this study the SGDNA has obtained from the General Archive of the Administration, several types of documents in relation to the February 28, 1969

Figure 8
Photograph showing the damage caused by the earthquake of February 28, 1969 inside the church of *Nuestra Señora de los Dolores* (Isla Cristina, Huelva) where the roof collapsed. Courtesy of the Municipal Archive of Isla Cristina

Figure 9
Photograph showing the damage caused by the earthquake of February 28, 1969 in the Cathedral of Seville where some of the Gothic ornaments were detached. It is one of the photographs taken on site by the CIFRA agency. Courtesy of the General Archive of the Administration

Table 7

Archival documents of February 28, 1969 earthquake

Document	Institution	Description
File of restoration of monuments of the *Dirección General de Bellas Artes* (03)115.000	*Archivo General de la Administración*	Restoration project of the ornaments of Seville cathedral
File of restoration of monuments of the *Dirección General de Bellas Artes* (03)115.000	*Archivo General de la Administración*	Memory of the restoration of the *Reales Alcázares* of Seville
Municipal report	*Archivo Municipal de Huelva*	List of buildings affected by the earthquake
Application for aid	*Archivo Municipal de Huelva*	Application for aid to repair damages caused by the earthquake. Set 2378
Municipal report	*Archivo Municipal de Isla Cristina (Huelva)*	Report of the technical architect with the list of affected homes
Municipal file	*Archivo Municipal de Gibraleón (Huelva)*	Subsidies expedient for the rehabilitation of homes for damages caused by the earthquake
Proceedings of the Municipal Permanent Commission, AHMB L-S 325	*Archivo Municipal de Badajoz (Extremadura)*	Description of how the earthquake was felt
Entry Registration of the Civil Government	*Archivo Histórico Provincial de Ciudad Real*	Record of a report of the Civil Guard indicating the downfall of the roof of a church, due to the earthquake
Municipal file of Faro (Portugal) nº PT/ MFAR/CMFAR/O-A/003/003 (1969-1977)	*Archivo Municipal de Faro (Portugal)*	Data on affected homes and requests for aid for damages caused by the earthquake

earthquake, and very useful sources of information for the study of the intensity of the earthquake.

• The restoration projects of specific monuments of the affected area and those of some significant public works carried out that year or later. The seismic causes that have led to this restoration are often indicated in the memory of such kind of projects.

• The memories of the Civil Governments. A chronicle of sorts each province sent to the

Table 8

List of newspaper clippings with information about February 28, 1969 earthquake

Newspaper	Date	Location
ABC	01/03/1969	National
ABC	01/03/1969	Madrid
ABC	02/03/1969	National
Nueva Alcarria	01/03/1969	Guadalajara
Flores y Abejas	04/03/1969	Guadalajara
Campo Soriano	01/03/1969	Soria
Soria Hogar y Pueblo	02/03/1969	Soria
Hoy de Extremadura	01/03/1969	Extremadura
Hoy de Extremadura	02/03/1969	Extremadura
Hoy de Extremadura	04/03/1969	Extremadura
Hoy de Extremadura	06/03/1969	Extremadura
Hoy de Extremadura	09/03/1969	Extremadura
Hoy de Extremadura	12/03/1969	Extremadura
Odiel de Huelva	01/03/1969	Huelva
Odiel de Huelva	02/03/1969	Huelva
Odiel de Huelva	04/03/1969	Huelva
Odiel de Huelva	30/04/1969	Huelva
Odiel de Huelva	19/05/1969	Huelva
Faro de Vigo	01/03/1969	Vigo
Faro de Vigo	04/03/1969	Vigo
Proa	01/03/1969	León
La Voz de Almeria	01/03/1969	Almería
La Nueva España	01/03/1969	National
Diario de Jaén	01/03/1969	Jaén
Diario de Jaén	05/03/1969	Jaén
Diario de Las Palmas	28/02/1969	Las Palmas de Gran Canaria
Diario de Las Palmas	01/02/1969	Las Palmas de Gran Canaria
Diario de Avisos	01/03/1969	Canarias
Madrid	28/02/1969	Madrid
Pueblo	01/03/1969	National
Pueblo	28/02/1969	National
Nuevo Diario	01/03/1969	National
Ya	01/03/1969	National
Alcázar	01/03/1969	Toledo
Nuevo Diario	01/03/1969	National
Informaciones	01/03/1969	National
Arriba	01/03/1969	National
Diario de Burgos	01/03/1969	Burgos
ABC Sevilla	01/03/1969	Seville
Diario de Cádiz	28/02/1969	Cádiz
Diario de Cádiz	01/03/1969	Cádiz
Diario de Cádiz	02/03/1969	Cádiz
Diario de Cádiz	05/03/1969	Cádiz
Diario de Cuenca	01/03/1969	Cuenca
Diario de Lisboa	28/02/1969	Lisbon, Portugal
Lanza de La Mancha	01/03/1969	Castilla - La Mancha
Línea	28/02/1969	Murcia
Línea	01/03/1969	Murcia

Table 8 *continued*

Newspaper	Date	Location
Norte de Castilla	01/03/1969	Castilla León
Valle de Elda	08/03/1969	Alicante
Mirador de San Fernando	03/03/1969	Cádiz
Revista QP Compañía Telefonica	01/01/1969	National
La Voz de Galicia	28/02/1969	Galicia
La Voz de Galicia	01/03/1969	Galicia
La Voz de Galicia	02/03/1969	Galicia
La Voz del Sur	01/03/1969	Málaga

Ministry of Government, covering the most outstanding aspects that occurred in the year.

- The Culture funds with photographic collections of the National Delegation of Propaganda and news clips depicting photographs of damage caused by the earthquake.
- The photographic collection of the CIFRA Agency.

Through Municipal Archives of the affected area, reports were collected from the Municipal Urbanism Technical Services where the damages of some houses are registered. In many cases, these reports include photographs.

Sometimes a Council shares information from its archives with the SGDNA: the proceedings of the Permanent Municipal Commission where the consequences of the earthquake are recorded, or the requests for aid for rehabilitation of houses damaged by the earthquake.

In many Provincial Historical Archives, the Civil Government fund is preserved wherever has been able to locate some references from the Entry registration of documents.

A list of the most relevant documents provided by these Archives is shown in Table 7.

4.7. Press News

This earthquake was widely felt throughout the Iberian Peninsula, Canary Islands and Morocco, so the printed press at this time echoed the news.

In the different Newspaper archives, Municipal archives and libraries it has been possible to find news clips reporting on how the earthquake was felt throughout the Spanish territory.

On the other hand, it has been more complicated to find news coverage of the event by the foreign press, with the exception of a Lisbon newspaper.

Table 8 shows a list of the press news clips from SGDNA covering the Saint Vincent Cape earthquake.

5. Conclusion

SGDNA has a wide and varied collection of documents useful for multiple studies in the field of seismology and geophysics, intended for users to access this information. With this in mind, a course of action has been taken, where:

- Efforts are being made to register the National Geophysical Data Archive in the Archives Guide Census of the Ministry of Culture and Sports. This Census is a directory of the archives of Spain and Latin America that allows citizens to immediately locate the archives centres as well as the funds and collections they safeguard and the services they provide.
- SGDNA is incorporated to the IGN website. The information—already available—is being prepared for its display in corporate website, a well-known place for researchers of the field.

Presently, the easiest way to contact SGDNA and request information for research use is via email: archivo.geofisico@fomento.es.

Publisher's Note   Springer Nature remains neutral with regard to jurisdictional claims in published maps and institutional affiliations.

REFERENCE

Martínez Solares, J. M. (2011). *Sismicidad pre-instrumental. Los grandes terremotos históricos en España*. Enseñanza de las ciencias de la Tierra, ISSN 1132-9157, Vol. 19, N°. 3, 2011.

(Received June 5, 2019, revised September 26, 2019, accepted October 9, 2019, Published online October 17, 2019)

Pure Appl. Geophys. 177 (2020), 1761–1780
© 2020 Springer Nature Switzerland AG
https://doi.org/10.1007/s00024-020-02475-3

Focal Parameters of Earthquakes Offshore Cape St. Vincent Using an Amphibious Network

ROBERTO CABIECES,[1] ELISA BUFORN,[2] SIMONE CESCA,[3] and ANTONIO PAZOS[1]

Abstract—Earthquakes with submarine foci are generally located with high uncertainties, and their focal mechanisms are poorly resolved, due to the confinement of the monitoring network onshore and the consequent poor azimuthal coverage. The use of amphibious networks, combining ocean-bottom seismometers (OBSs) and land stations, helps to reduce the epicentral distance and the azimuth gap, thus better constraining the hypocentral locations and decreasing the focal mechanism uncertainties. A second important factor in improving the location accuracy for offshore seismicity is the use of a suitable velocity model. The objective of this paper is to study how the combination of an amphibious network with 3D crustal models can improve offshore earthquake hypocenter locations and focal mechanisms. The study area is SW Iberia near Cape St. Vincent, which generated some of the most striking earthquakes and tsunamis in Europe in past centuries, such as the 1755 Lisbon earthquake ($I_{max} = X$). We deployed an array of six broadband OBSs 200 km offshore Cape St. Vincent to study the seismicity of the region for a period of 8 months. During this period, we detected 52 earthquakes, the largest with magnitude M (mbLg) ≈ 5. Thirty-eight earthquakes are relocated using land stations in Iberia and North Africa and the OBS array with different velocity models. Focal mechanisms and moment tensors are computed for a data set of seven earthquakes based on first-motion polarities, body-wave amplitude spectra and waveform cross-correlation. We show that if we include offshore OBS data and consider accurate 3D velocity models to locate earthquakes, the focal parameter uncertainties decrease substantially, thus improving the depth constraint. We also show that the locations and focal mechanisms obtained using the amphibious network agree with the regional stress pattern in the SW Iberia region.

Keywords: Earthquakes, Cape St. Vincent, hypocentral relocation, amphibious network, 3D velocity models, focal mechanisms.

Electronic supplementary material The online version of this article (https://doi.org/10.1007/s00024-020-02475-3) contains supplementary material, which is available to authorized users.

[1] Royal Spanish Navy Observatory, Cecilio Pujazon s/n, 11100 San Fernando, Spain. E-mail: rcabdia@roa.es
[2] Fac. CC. Físicas, Universidad Complutense de Madrid, 28040 Madrid, Spain.
[3] GFZ German Research Centre for Geosciences, Potsdam, Germany.

1. Introduction

The southwestern Ibero-Maghrebian region (i.e., from 14° W to 5° W and 34° N to 39° N) is a complex seismotectonic region (Fig. 1) located on the border between the western part of the Eurasian and African plate boundary which extends from the Azores islands to the Ibero-Maghrebian region (Iberian Peninsula and northern Africa). Their convergence is evolving at low rates of 3–5 mm/year (Fernandes et al. 2003; Serpelloni et al. 2007; Nocquet 2012) with a general NW–SE orientation (McKenzie 1972; Udias et al. 1976), and the tectonic compressional stress is accommodated dominantly by thrust-faulting earthquakes with NW–SE-oriented pressure axes (Buforn et al. 1988a, b, 2004; Custódio et al. 2016).

Two of the largest and most destructive earthquakes that occurred in Europe, i.e., the 1755 Lisbon earthquake ($I_{max} = X$; Martínez Solares and Arroyo 2004; Baptista and Miranda 2009) and the 1969 earthquake (Ms = 8) (López Arroyo and Udias 1972; Fukao 1973; Buforn et al. 1988a), were located offshore Cape St. Vincent (CSV). The spatial distribution of the seismicity in this region is diffuse (Buforn et al. 1988a, b; Sartori et al. 1994; Bezzeghoud and Buforn 1999; Zitellini et al. 2009; Martínez-Loriente et al. 2013), with several clusters of small-magnitude earthquakes highlighting the high number of faults and lineations with different orientations (Buforn et al. 2004, 2019; Stich et al. 2020). Figure 1 shows the seismicity for the studied area (period 2014–2019) taken from the Instituto Geográfico Nacional (IGN). In general, west of the Gibraltar Arc, the hypocenters have large depth errors due to the poor azimuthal coverage. Seismicity mostly occurs at shallow depths (h < 40 km, red circles in Fig. 1), but intermediate-depth seismicity

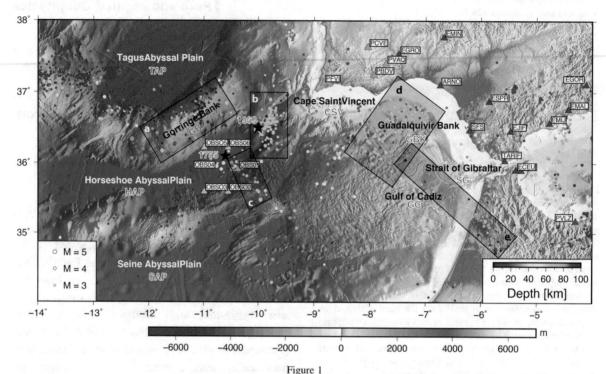

Figure 1
OBS array, land stations and hypocenter distribution from 2010–2015 for earthquakes with magnitude greater than 3 (Catalogue of Instituto Geográfico Nacional, Madrid http://www.ign.es/web/ign/portal/sis-catalogo-terremotos). The shaded squares *a*, *b*, *c*, *d* highlight the most remarkable hypocenter clusters and the alienation in the square area *e*. The color scale corresponds to the earthquake depth, and the size of the circles are proportional to the earthquake magnitude. Stars correspond to the 1755 Lisbon earthquake (I_{max} = VIII) and 1968 earthquake (M_s = 8.0). Green triangles = OBS array, orange triangles = IPMA network, red triangles = WM network, blue triangles = IGN network. *CSV* Cape St. Vincent, *HAP* Horseshoe Abyssal Plain, *TAP* Tagus Abyssal Plain, *SAP* Seine Abyssal Plain, *GC* Gulf of Cadiz. General Bathymetric Chart of the Oceans (GEBCO) digital atlas (http://www.gebco.net/)

(40 km < h < 130 km) is also present, being most active east Gibraltar Arc (Fig. 1). Onshore, seismicity occurs in both southern Iberia and northern Morocco, and depths tend to increase when approaching the Iberian Peninsula (red to yellow-green circle pattern in Fig. 1).

From Fig. 1, we observe that west of 12° W (from 36° to 38° N), epicenter clusters seem to sketch an E–W lineament pattern. However, the distribution of epicenters is more complicated to the east, where several clusters of earthquakes show different orientations. The first cluster (cluster *a*) in Fig. 1 is located at ∼ 37°N, with epicenters distributed in a NE–SW direction along the Gorringe Bank (GB). Clusters at ∼ 37° N and 10° W (cluster *b*) and aligned in the SE–NW direction (Fig. 1), and at approximately ∼ 36° N and 10°–11° W and aligned in the N–S direction (cluster *c*) (Fig. 1), host the locations of the

1969 earthquake (cluster *c*) and the proposed hypocenter for the 1755 earthquake (left top corner of cluster *b*) (black star, Martínez Solares and Arroyo 2004), respectively. A fourth cluster (cluster *d*) is located in the Gulf of Cadiz (GC), from ∼ 8° W to ∼ 6.5° W, distributed in the NE–SW direction (Fig. 1), with a secondary alignment (cluster *e*) in the NW–SE direction (Fig. 1) reaching the Moroccan coast (see Buforn et al. (2019) for further details). Most of these clusters have been identified, and their general patterns have been discussed by Custódio et al. (2016).

Earthquakes offshore CSV are generally located with high uncertainties, due to both the poor azimuthal coverage and the lack of seismic stations near the sources. These complications have been partially overcome by temporary ocean-bottom seismometer (OBS) deployment experiments in the region

(Geissler et al. 2010; Grevemeyer et al. 2016, 2017; Silva et al. 2017). Another problem affecting an accurate hypocentral location is the complexity of the crustal structure in the region. Large changes in crustal thickness have been suggested for this region (Palomeras et al. 2008; Sallarès et al. 2011). In addition to the aforementioned complex geodynamic context and the intricate fault systems (Sallarès et al. 2013; Martínez-Loriente et al. 2013, 2018), long refraction and wide-angle profiles identify sediment thicknesses ranging from 3 to 8 km offshore the Gibraltar Strait (Sallarès et al. 2011). These sediments might affect seismic wave propagation along different directions (Custodio et al. 2012), strongly attenuating the seismic waves and thus hindering the waveform records in the ocean floor. Such complex crustal structure and lateral heterogeneities affect earthquake relocation when using simplified 1D crustal models. The location accuracy can be improved using high-resolution 3D velocity models (e.g., velocity grid cells with a size of $\sim$ 1 km), which provide more reliable P and S estimated travel times used in earthquake location algorithms. The location improvement is especially significant for regions affected by sharp variations in the lithosphere rheology properties and in the presence of sparse seismic networks.

In Fig. 1, we show the OBS array and the distribution of the onshore seismic stations used in this work. The array had a configuration with an $\sim$ 40 km average interstation distance and 80 km aperture. Unfortunately, one station, OBS4, was lost. This paper discusses the seismological analysis of OBS data in combination with onshore records from the closest stations, to detect, locate and characterize weak to moderate seismicity during the time of the OBS array deployment. We aim to assess the improvement in seismicity characterization by comparing the location procedure using the amphibious network with the location results based on land stations only. We further discuss the improvement in location results based on different 1D and two different 3D crustal models. A main goal of this work is to assess the benefit of seismicity monitoring offshore CSV, using a combination of onshore and offshore seismic stations and accurate 3D velocity models. Furthermore, focal mechanisms and moment tensors

are determined using land stations and OBSs for selected earthquakes. The results of this study are discussed and compared with solutions from the IGN catalog and previous works that also used amphibious networks in the same region.

2. Data and Methods

2.1. Data

From September 2015 to April 2016, on behalf of the ALERTES-RIM project, the Spanish Navy deployed the six OBSs (Supplementary Material, Table 1) of the FOMAR (https://doi.org/10.14470/jz581150, ROA/UCM) in the Horseshoe Abyssal Plain (HAP) (Fig. 1) at 4 km water depth. OBS01, OBS02 and OBS03 were each equipped with a three-component broadband seismometer (Güralp CMG-40T, 60 s–50 Hz), and OBS04 (lost), OBS05 and OBS06 with a Trillium Compact (120 s–100 Hz; Nanometrics). All of the OBSs had the same hydrophones (HTI-04-PCA/ULF, 100 s–8 kHz; High Tech, Inc.), with the sampling frequency established at 50 samples/s. The response of the seismometer and hydrophone were built from the MCS data logger and the sensor specifications (CMG-40T and Trillium Compact). We show the responses of these sensors in the supplementary material, Fig. A1. The raw seismograms (Fig. 2a) were deconvolved to ground motion (Fig. 2b) and velocity (Fig. 2c) and filtered with a zero-phase high-pass filter with a corner frequency of 0.5 Hz. In this way, ocean-induced microseisms, which are dominant in the frequency band [0.05–2] Hz (Webb 1998), were partially removed, facilitating a better estimation of the arrival times (Fig. 2d–f).

An additional difficulty related to the OBS records is to know the real positions of the OBSs on the sea bottom and the correct orientations of the horizontal components and to have the OBS clocks synchronized with the rest of the station network. The OBS positions were estimated using active acoustic techniques (Shiobara et al. 1997), with an uncertainty of hundreds of meters. The timing errors were corrected (Cabieces et al. 2018, 2020) by using the daily differences between daily noise cross-correlation

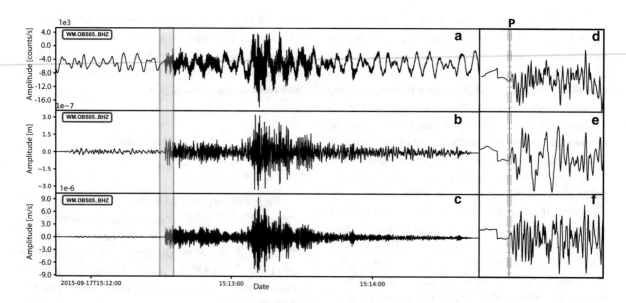

Figure 2

a Raw seismogram from OBS05 (earthquake T1, Supplementary Material Table 2, magnitude mb 4.9), with an epicentral distance of 400 km from the OBS array and 57 km depth. **b** Seismogram in displacement. **c** Seismogram in velocity. **b**, **c** The seismogram is filtered with a high-pass filter (fc = 0.5 Hz)

Green's functions (NCCGF) and a reference created from the stack of all NCCGF (Stehly et al. 2007; Sens-Schönfelder 2008) following the time symmetry analysis algorithm (Gouedard et al. 2014). The vertical component was realigned by a gimbaling system (Stähler et al. 2016), and the horizontal components of every OBS were oriented (Cabieces et al. 2020) by correlating Rayleigh waves recorded in the radial and vertical components from shallow teleseismic earthquakes (Stachnik et al. 2012). For the signal amplitudes, we compared displacement waveforms (after deconvolving the instrument transfer function) at the OBS stations and at the closest land stations for a few teleseismic events. Synthetic P and S waveforms reproduced observed waveforms well, except for the signal amplitudes, which were different. We attributed this discrepancy to the unknown coupling of the OBS sensors on the seafloor. Therefore, in the following source inversion section, we will ignore the signal amplitude.

2.2. Velocity models

We consider three Earth seismic velocity models for the CSV region (14° W to 5° W and 34° N to 38° N) to locate earthquakes. The first model, later referred to as the 1D-Iberia model (Martinez Solares 1995), consists of a crust formed by three flat horizontal layers of constant velocity for a total depth of 31 km over a mantle (see Fig. A4, supplementary material). In addition, we use two 3D crustal models that cover the same geographic area (14° W to 5° W and 34° N to 39° N). The first 3D model, proposed by Grandin et al. (2007), is here referred to as the 3D-1 model. It is defined for cells of 1 × 1 × 1 km and reaches a maximum depth at 60 km. The second 3D model, 3D-2 model, was proposed by Civiero et al. (2018). Cubic cells have dimensions of 10 × 10 × 0.5 km but reach a deeper maximum depth of 250 km (see Fig. A4 and the video in the supplementary material for further details). For the determination of focal mechanisms with first polarities, take-off angles are estimated using a crustal model with a linear velocity gradient. We advocate a gradient a gradient velocity model to estimating the focal mechanism with the 1D-Iberia, because it provides a more continuous distribution of take-off angles. The gradient velocity model used in this work has been widely applied in the SW part of the Ibero-Maghrebian region (Buforn et al. 1988a, b).

For the moment tensor inversion, we utilize the 1D-Iberia model.

2.3. Earthquake Detection

OBS and land station data (from onshore broadband stations of the WM (Western Mediterranean), IGN (Instituto Geográfico Nacional) and IPMA (Instituto Português do Mar e da Atmosfera) seismic networks) were scanned to search for local earthquakes. Preliminary detection was performed using SeisComP3 (http://www.seiscomp3.org), imposing a minimum of six stations (OBSs or land stations) with a signal-to-noise ratio larger than 4. We applied a short-time-average over long-time-average (STA/LTA) in a previously filtered seismogram (high-pass filter fc = 0.5 Hz) to automatically detect events. The STA was set to 1 s, while the LTA was set to 40 s. We detected a total of 52 earthquakes; in this paper, we only discuss a subset of 38 earthquakes, all of which were detected by all OBS records and at least one land station. Two earthquakes were located using all OBSs and a few onshore stations and do not appear in the IGN catalog (https://www.ign.es/web/ign/portal/sis-catalogo-terremotos).

2.4. Hypocentral Location

The hypocenter location carried out by the IGN, which is used for comparison, uses the linear algorithm LocSAT (Bratt and Bache 1988), based on the minimization of the residuals between theoretical and observed arrival times through the generalized inverse matrix of an a priori model. It starts the location procedure assuming a trial location near the station that first detects a seismic phase. It then calculates theoretical arrival times in accordance with a 1D velocity model (1D-Iberia model) and the residuals between theoretical and observed arrival times. The hypocentral locations are then iteratively improved to minimize the residuals. The residuals between the theoretical and observed arrival times at different stations are estimated as follows:

$$r = A\Delta x \tag{1}$$

where r is the residual vector, A is the model matrix of partial derivatives of travel time with respect to the spatial hypocenter coordinates at the current trial location, and Δx is the perturbation vector. The solution to this equation is to estimate r through the least mean squares method:

$$\Delta x = (A^T A)^{-1} A r \tag{2}$$

Here, the $(A^T A)^{-1} A$ is the *generalized inverse* of A, computed by singular-value decomposition. Now, the new trial is $x_c + \Delta x$ and it can iterate the process up to find a required minimum residual r. The confidence ellipsoid is derived following Jordan and Sverdrup (1981) and Bratt and Bache (1988) from the points x_e in a p percent confidence ellipsoid for the solution x,

$$(x_e - x)^T V_x^{-1} (x_e - x) = k_p^2 \tag{3}$$

where $V_x = (A^T A)^{-1}$ and k_p is the confidence coefficient.

In our study, the hypocentral estimations are carried out using the nonlinear hypocenter location method NonLinLoc (NLL, Lomax et al. 2001), which has the ability to relocate earthquakes using 1D or 3D velocity models. NLL relies on a probabilistic approach to perform the earthquake location (Tarantola and Valette, 1982). The nonlinear global search location can be obtained using a systematic grid search, a stochastic Metropolis–Gibbs sampling approach (Sambridge and Mosegaard 2002) or the sampling algorithm Oct-tree (Lomax and Curtis 2001).

NLL allows us to compute a complete probabilistic solution through the knowledge of an a priori probability density function (PDF) (bounds of the region in which we know the event occurred) and the evaluation of the equal differential time (EDT) likelihood function (Zhou 1994; Font et al. 2004; Lomax 2005) given by

$$L(\boldsymbol{x}) = \left[\sum_{a,b} \frac{1}{\sqrt{\sigma_a^2 + \sigma_b^2}} e^{\left(\frac{-\left\{ \left[T_a^O - T_b^O \right] - \left[TT_a^c(x) - TT_b^c(x) \right] \right\}^2}{\sigma_a^2 + \sigma_b^2} \right)} \right]^N \tag{4}$$

where x values are the spatial hypocenter trial coordinates (x_o, y_o, z_o), T_a^O and T_b^O are the observed arrival times, TT_a^c and TT_b^c are the calculated travel times for two observations a and b, σ_a and σ_b are the standard deviations of every pair of observations and N is the total number of observations.

The travel times between a station and all nodes of the 3D grid are calculated using the eikonal finite difference scheme (Podvin and Lecomte 1991), with the restriction of cubic grid cells. Additionally, NLL uses the gradients of the travel times at the nodes to estimate the take-off angles. Finally, the posteriori PDF is the product of the a priori information and the likelihood function. In this paper, we use Oct-tree as the sampling algorithm (integrated into the NLL package), which is shown to be 100 times faster than the conventional grid search (Lomax and Curtis 2001) and derives a PDF comparable to that from the Metropolis–Gibbs sampling approach. Oct-tree uses recursive subdivision and sampling of rectangular cells in three-dimensional space to generate a cascade structure of sampled cells (Lomax 2008). The Oct-tree algorithm is adequate for complicated or multi-modal PDFs because its recursive subdivision procedure converges rapidly and robustly in intricate 3D velocity models (Husen et al. 2003).

The uncertainties in latitude, longitude and depth are estimated from the square root of the eigenvalues of the covariance matrix associated with the earthquake location. In the following sections, to compare the LocSAT and NLL earthquake locations, we transform the NLL confidence ellipse and depth error estimated with a 68.3% uncertainty level to a common uncertainty level of 90.0%.

2.5. Magnitude Estimation

We follow López (2008) to estimate the local magnitude (mbLg) applying the equation

$$mbLg = \log\left(\frac{A}{T}\right) + 1.17\log(\Delta) + 0.0012\Delta + 0.67$$

where A is the amplitude measured in (μm) and T is the period of the Lg phase, measured at the horizontal component, and Δ is the epicentral distance. This magnitude is published in the IGN catalog and is the reference for the majority of the locations carried out in this work. Because of the limitations of this magnitude mbLg (i.e., distance from Iberia < 200 km

and depth < 50 km), for some earthquakes of the data set, we calculate the magnitude mb.

2.6. Focal Mechanism Determination

We independently use first-motion polarities to determine pure double couple (DC) focal mechanisms and a waveform-based moment tensor (MT) inversion to determine a deviatoric moment tensor. For the DC determination, we follow the approach by Brillinger et al. (1980), which evaluates the maximum likelihood function to correlate the stress orientation with the theoretical amplitude of the P-wave and the observed polarities. We estimate the P–T axes, fault-plane errors and score as quality parameters. The score is defined as the number of correct polarities over the total number of polarities (Brillinger et al. 1980). The use of a gradient velocity model to determine the focal mechanism permits us to obtain a more continuous distribution of take-off angles than using 1D models, such as the 1D model Iberia. However, we do not account for the significant differences between take-off angles estimated using the gradient velocity model and using 1D model Iberia for the earthquake data set in this work.

The second approach is based on a probabilistic MT optimization (Grond, Heimann et al. 2018). In our implementation, we fit P and S waveform amplitude spectra in the frequency band 0.5–3.0 Hz at onshore stations, up to a distance of 350 km. Additionally, we fit the cross-correlations of P and S waveforms at all stations, including OBSs, up to a distance of 350 km. In this way, we can use the reliable waveform information at OBSs without modeling the signal amplitude, which cannot be easily reproduced (see Sect. 2.1). P and S waveforms are tapered, using time windows of 6 and 8 s and starting 2 and 3 s before the theoretical phase arrival time, respectively. We apply a bandpass filter between 0.5 and 3.0 Hz (0.5–2.5 Hz for selected earthquakes) to maximize the signal-to-noise ratio. Grond is set up to simultaneously determine centroid location and depth, centroid time, scalar moment (and thus magnitude) and MT components (here for a deviatoric MT). We limit the depth search according to the depth estimates and uncertainties obtained after the hypocentral location. We use the 1D-Iberia model to model synthetic seismograms for P and S phases. Grond provides a solution using all available data, but simultaneously performs a bootstrap analysis, obtaining an ensemble of MTs for different data configurations, which are assessed to produce a 'mean' MT solution. Typically, stable inversions lead to similar mean and best solutions. In the present case, given the low magnitudes, challenging high-frequency fit and noise contamination, we discuss only those seven events for which the best and mean solutions are in agreement. For the MT inversion, we used a combination of amplitude spectra fit and cross-correlation, taking into account the location uncertainty. Basically, the MT geometry is resolved by a combination of amplitude spectra and waveform shape fits, using all available data and onshore/offshore stations, whereas the scalar moment (and consequently the magnitude) is resolved from amplitude spectra at land stations only. Synthetic waveform and spectra were computed using the 1D-Iberia crustal model. Since the local crustal structure at the offshore location is poorly known, we verified the stability of the MT inversion for a selected earthquake (T1), using a modified velocity model (see supplementary material and Fig. A3). Results show that synthetic seismograms, even for body-wave phases, are poorly affected by changes in the shallow part of the crustal structure, probably because of the steep incidence angles. As a consequence, the MT inversion is stable, producing only minor changes in the focal mechanism and depth estimations. To our knowledge, this is a pioneer attempt to jointly perform an MT inversion for moderate seismicity offshore through the joint modeling of land and OBSs data.

3. Results

3.1. Hypocentral Determinations

In this section, we summarize the location results for the earthquake data set (38 earthquakes, magnitudes $M \geq 2.5$) that we obtained using the three previously described Earth velocity models, onshore and offshore data, and the common location algorithm (NLL). P- and S-phase picking is performed for all five OBSs and combined with these phases recorded at land stations with epicentral distances shorter than 700 km, assuming a flat Earth approximation. The hypocentral locations using the 1D-Iberia, 3D-1 and 3D-2 models are compared with those obtained by the IGN (Supplementary Material, Tables 2–5). Note that the IGN locations are also estimated using the 1D-Iberia model and are named IGN. Figure 3 shows a comparison of location results (left) and uncertainty ellipses (right) between the IGN locations and the locations in this work (1D-Iberia, 3D-1 and 3D-2 from top to bottom, respectively). We define three spatial domains from west to east (Fig. 3) between 34° and 38° N. Region A is located west of the OBS array (west of 11° W) corresponding roughly to the Horseshoe Abyssal Plain, region B lies between the western HAP and the Gulf of Cadiz (8°–11° W), and region C is located around the Strait of Gibraltar (5°–8° W). The selection of these regions is not based on changes in the seismicity patterns but rather distinguished among regions with different locations and focal mechanism resolutions (i.e., region A is west of the array; region B, between array and land stations; and region C, close to land stations).

Figure 3a, d compares the hypocentral locations by the IGN (black symbols and gray ellipses) with those using 1D-Iberia (orange symbols and ellipses). In addition to the different algorithms used for the locations, the comparison basically illustrates the benefits of including OBSs. Note that circle and square symbols in Fig. 3 denote shallow (< 40 km depth) and intermediate-depth (> 40 km) hypocenters, respectively, and their sizes are proportional to their magnitudes. The locations appear in relative agreement, but differences in epicentral location and depth become progressively larger proceeding westward. In detail, we observe relatively large differences in region A, where the contribution of the OBSs is negligible because the OBSs array does not contribute to closing the large azimuthal gap. In region B, the OBSs help to constrain the depths, which appear in many cases deeper than the reference depths, and for most of the events, the locations are shifted to the SW ∼ 20 km and better align along the NW–SE direction. Finally, in region C (the OBS closest to the region C earthquakes is farther than

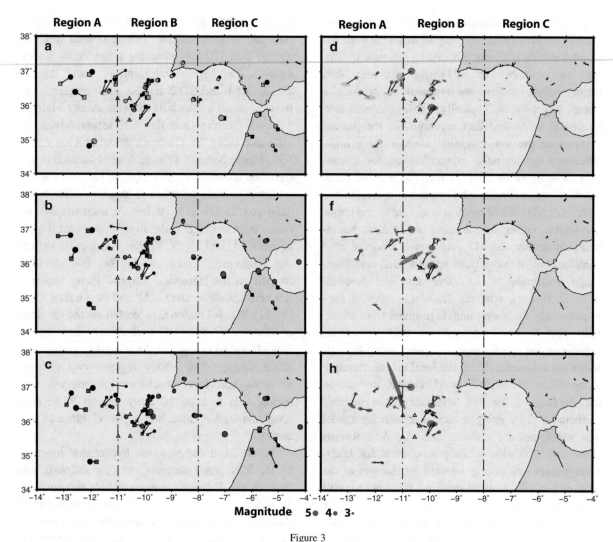

Figure 3
Hypocenter distribution and confidence ellipses associated to the locations. **a** Earthquake hypocenter location, comparison between IGN (black color) and 1D-Iberia (orange color); circles, hypocenters above 40 km; squares, hypocenters below 40 km. **b** Earthquake hypocenter location, comparison between IGN and 3D-1. **c** Same as **b** and **a** but comparison between IGN and 3D-2. **d** Confidence ellipses (90% level) associated to the epicenters IGN (black color) and 1D-Iberia (orange color). **e** Confidence ellipses (90% level) associated to the epicenters IGN (black color) and 3D-1 (red color). **f** Confidence ellipses (90% level) associated to the epicenters IGN (black color) and 3D-2 (blue color)

250 km), we do not observe high discrepancies between the locations from the 1D-Iberia and 3D models or the IGN, either in epicentral locations or depths, although 3D-1 and 3D-1 suggest deeper sources than the IGN results. A comparison of location uncertainties (Fig. 3d) shows that these are reduced in region B, where the azimuthal coverage is substantially improved when including the OBSs. Here, the average maximum and minimum semi-axes of the uncertainty ellipses decrease slightly from

~ 10 and ~ 9 km (reference IGN locations) to ~ 10 and ~ 5 km (1D-Iberia). In region C, the location uncertainties between the IGN and our locations are comparable, and the major and minor semi-axes of the uncertainty ellipse also decrease from ~ 6 and ~ 3 km to ~ 6 and ~ 2 km, respectively.

Figure 3b, e (3D-1 red symbols and red ellipses) illustrates the location improvement when the 3D model is considered. In comparison with the IGN catalog, these locations present features similar to those discussed for 1D-Iberia: a general agreement in locations, decreasing differences in epicentral location and depth towards the east (near the small OBS network), and deeper locations in region A and in parts of region B. The most remarkable difference is the reduction in the location uncertainty by $\sim 50\%$ in all regions, which was quantified by comparing the average size of the uncertainty ellipse axis (see the quantified parameters in Sect. 4, Fig. 9).

The last comparison (Fig. 3c, h) concerns the second tested 3D model, 3D-2, whose locations and uncertainties are plotted in blue. A comparison with the IGN solutions confirms previous major general findings. A comparison with 3D-1 shows, in general, minor differences: the two models predict similar locations and comparable uncertainties in all three domains.

Figure 4 shows a vertical W–E hypocenter cross-section (5°–14° W) through the three domains A, B and C, showing depth error bars. The highest depth errors occur in region A independent of the model used for the location. However, the most remarkable finding is that the IGN locations (Fig. 4d) have fixed the depths, which is necessary in the implementation of LocSAT to estimate the epicenters. In region B, the depth errors are greatly reduced for 3D-1 (Fig. 4b) and 3D-2 (Fig. 4c), and finally, region C, the region with the minimum difference between the IGN and the rest of the locations, is shown.

Two earthquakes (T22 and T30) in region A show special location problems related to the 3D models and/or with outliers in travel time estimation; a detailed analysis is given in Sect. 3.2.

3.2. Earthquakes T22 and T30: Challenging Locations Outside the Network

The results obtained with model 3D-1 for event T22 relocation exemplify the difficulties in obtaining an accurate relocation using a limited number of stations placed on one side of the event (in this case, only OBS data are available). The large gap in the locations corrupts the performance of the algorithm. Figure 5 shows the PDF for T22 from 3D-1, which has a vertically elongated shape and multiple local maxima with a broad range of samples, spanning over

~ 50 km both laterally and vertically. The elongation of the PDF can be caused by the lack of good-quality S-wave arrival times (mean residual value of 0.3 s) and/or by the geometry of the OBS array. In the latter case, the seismic rays may leave the source with similar dip angles, leading to a strong trade-off between origin times and depth, thus leading to a lack of depth constraint. The most robust location achieved for the same event using model 3D-2 may be attributed to the higher depth resolution of this velocity model (dz = 1 km for 3D-1 vs. dz = 0.5 km for 3D-2).

Another interesting case study is earthquake T30 (Fig. 6). This earthquake shows relocation problems with model 3D-2. The PDF is very elongated in latitude (~ 50 km), but the depth is relatively well constrained (Fig. 6d, depth uncertainties of 22 km, Supplementary Material Table 4). Note that this earthquake is also located with model 3D-1 and has much lower uncertainties (~ 27 km in latitude, ~ 18 km in depth) using the same arrival times. For this reason, the large size of the confidence ellipsoid using model 3D-2 cannot be attributed to the network geometry but to the velocity model. Our interpretation is that since this earthquake occurs near the GB, the EDT likelihood function becomes unstable in the presence of sharp velocity changes in the uppermost model cells. An example of lateral velocity discontinuities can be found between the GB and the HAP (see the video in the supplementary material for further details).

3.3. Focal Mechanisms

We determine the DC focal mechanisms, using first-motion polarities and deviatoric MTs, for seven events of the data set with clear P arrivals and a minimum of ten records corresponding to OBS or inland stations. The results are summarized in Fig. A2 and Tables 6 and 7 in the supplementary material. Figure 7 displays the focal mechanisms estimated in this study, comparing our analysis with focal mechanisms of large earthquakes ($5.3 < M < 8.0$ from 1960 to 2009) obtained in previous studies (Buforn et al. 2004; Pro et al. 2013).

The seven selected earthquakes are located in regions B and C, from the GB to the GC and the

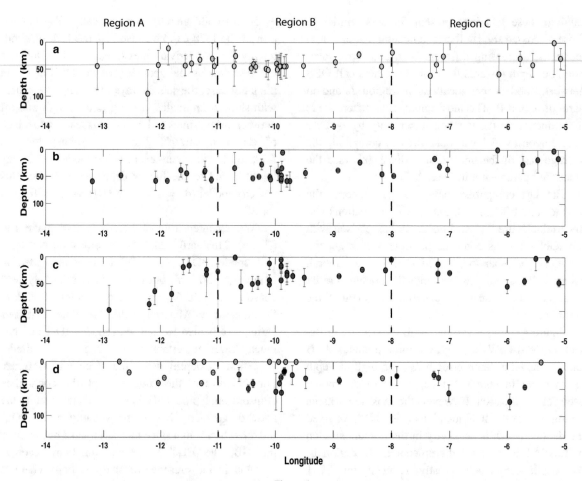

Figure 4

W–E profile showing earthquake depth and error bar estimated 90% confidence level. Circles (orange, red and grey), hypocenter with estimated depth error; green circles, hypocenter with fixed depth (http://www.ign.es/web/ign/portal/sis-catalogo-terremotos). **a** Depth earthquake and error bars from 1D-Iberia. **b** Depth earthquake and error bars from 3D-1. **c** Depth earthquake and error bars from 3D-2. **d** Depth earthquake and error bars from IGN

Strait of Gibraltar. Focal mechanisms are mostly characterized by thrust to strike-slip mechanisms (only T17 shows a normal oblique mechanism). All polarities and MT solutions, with the exception of T28, have a horizontal pressure axis oriented in the NNW–SSE direction following the regional stress pattern derived from the Eurasian–Nubian convergence. Fault plane solutions are poorly constrained due to a low number of polarities (from 11 to 19) and sparse azimuthal coverage. In spite of this difficulty, our solutions are in agreement to this regional stress pattern with predominant thrust solutions in CSV and GC (events T34, T28, T33 and T18), strike slip

solution near the Strait of Gibraltar (T1 and T11) and a normal solution in south Spain (T17).

MT solutions are shown in Fig. 7, where they can be compared to focal mechanism solutions (see Fig A2, supplementary material). For almost all cases, with the exception of event T28, we obtain a horizontal pressure axis oriented NNW–SSE to N–S. A larger discrepancy is observed for event T28, where the strike-slip motion of the polarity solution follows this stress pattern orientation, but the MT solution does not. These discrepancies for event T28 may be explained by the poor azimuthal coverage of observations. Examples of waveform and spectra fits, MT solutions and decompositions and Hudson plots

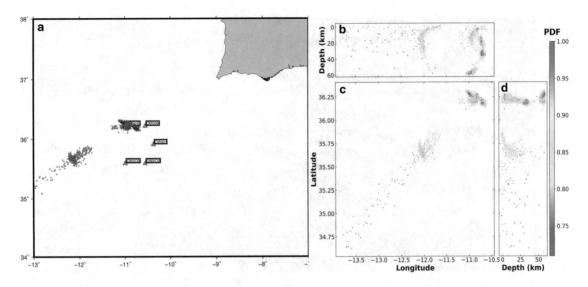

Figure 5
Relocation for earthquake T22, 3D-1. **a** PDF in the geographic projection. Detailed PDF, **b** longitude–depth, **c** longitude–latitude, **d** depth–latitude

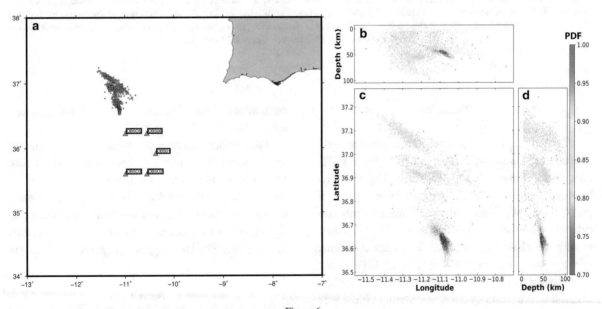

Figure 6
Relocation for earthquake T30, 3D-1. **a** PDF in the geographic projection. Detailed PDF, **b** longitude–depth, **c** longitude–latitude, **d** depth–latitude

(Hudson et al. 1989) are shown in Fig. 8. Two earthquakes have large amounts of non-DC (CLVD): events T11 (45%) and T18 (63%). Nevertheless, for T18, solutions obtained from first-motion and MT inversion are practically the same and are in agreement with the 1964 earthquake ($M_s = 6.4$, Buforn et al. 1988a, b).

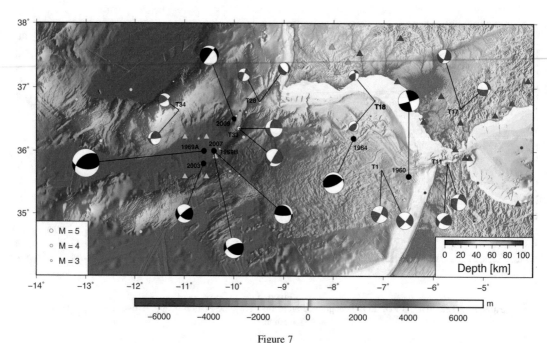

Figure 7

Focal mechanisms obtained in this study (in red polarities solution, Table 1, and in blue moment tensor inversion, Table 2). Epicentral locations 3D-1 with the depth in color scale and magnitude proportional to the size. In black, focal mechanisms of the largest earthquakes ($5.3 < M < 8.0$ from 1960 to 2009) in this region (Pro et al. 2013). Green triangles = OBS array, orange triangles = IPMA network, red triangles = WM network, blue triangles = IGN network

4. Discussion

First, we compare and analyze the most remarkable differences among focal parameters (mean RMS, uncertainty ellipse area and depth error, Fig. 9) for all locations in relation to the geographic regions A, B and C. Next, we analyze the focal mechanisms calculated with first polarities and MT inversion. Finally, we discuss our location with respect to other relevant works that use OBS data in the CSV area.

In region A (Fig. 9a, d, g), location 3D-1 has the smallest errors (RMS of 0.3 s, uncertainty ellipse area of 200 km² and depth error of 15 km) among all parameters, and the IGN has the largest errors. In region B (Fig. 9b, e, h), all of the focal parameter uncertainties decrease in comparison with region A, having similar 3D-1 and 3D-2 mean values for all parameters, but 1D-Iberia and the IGN have up to 50% higher uncertainties than locations using 3D models. In region C, Fig. 9c, f, i shows that 3D-1 decreases the uncertainty in all parameters in

comparison with 1D-Iberia and the IGN, and has errors similar to those of 3D-2.

Further details about the location quality and its uncertainties show that hypocenters in region A are generally poorly constrained, with large errors in latitude, longitude and depth. In this region, the error ellipses commonly follow a west–east orientation, so that epicenters are better constrained in latitude than in longitude. The large hypocenter parameter location

Figure 8

a Comparison of cross-correlated observed (black lines) and synthetic (red lines) displacement waveforms for selected station and components (the gray filled line represents the misfit for different time samples); each panel lists the station name, epicentral distance and azimuth. **b** Left: deviatoric moment tensors and their standard decomposition into double couple (DC) and compensated linear vector dipole (CLVD) are shown for the best solution using all data and for the mean solutions after data bootstrapping; right: best moment tensor solutions are represented in a Hudson diagram; the best solution of the left panel is identified by a square, the mean one by a large focal sphere. For comparison with the first-motion polarity, see Fig. A2 in the supplementary material

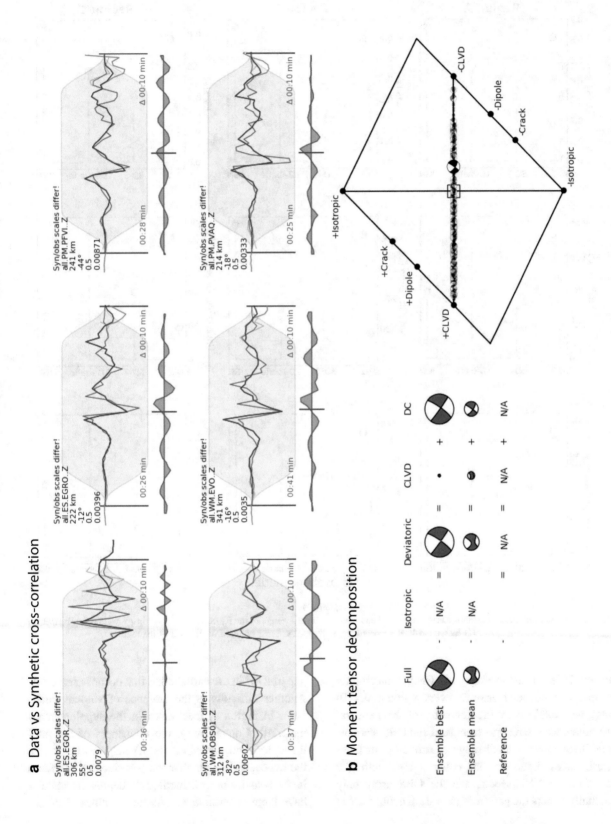

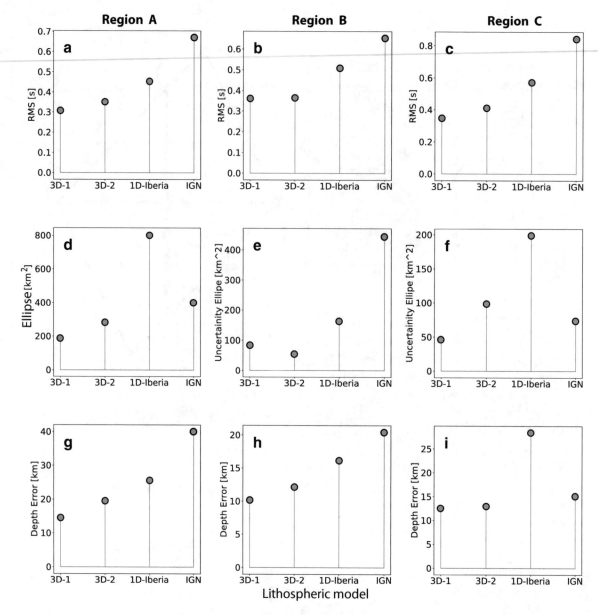

Figure 9
Average location parameters vs. lithospheric model. **a**, **d** and **e** represent the average of the RMS, uncertainty ellipse area and depth error for 3D-1, 3D-2, 1D-Iberia and IGN, respectively, in region A. Same notation for the rest of the regions

uncertainties (see average of RMS, uncertainty ellipse area and depth error for every region in Fig. 9) may be explained by the geometry of the network. Although we use records from land and OBS stations, the OBSs cover a small range of azimuths for these earthquakes. Land stations cover the azimuth gap only in the NNE sector, and the OBS array only partially covers the gap from the east, leaving a mean

gap of 270° in the earthquakes that occur in region A. Another difficulty is the presence of velocity interfaces in region A (see video in the supplementary material). Conveniently, the robustness of the non-linear location algorithm (NLL) helps to determine the hypocenter depth near sharp horizontal interfaces in the velocity model. Earthquake depths in region A have large uncertainties, with mean values of 26, 15 and 20 km, using models 1D-Iberia, 3D-1 and 3D-2 (Fig. 9a–c, respectively), but the IGN locations, which use LocSAT, have the depth fixed to be able to determine the epicenter. In summary, in region A, the use of a 3D model is the most important factor for improving the locations, rather than the use of the OBS array. This result occurs because 3D models reproduce the oceanic crust and upper mantle more reliably than the 1D model and thus can more accurately estimate the travel times between the oceanic and the continental lithosphere.

In region B, two results are the most remarkable. The first is that the epicenters near the OBS array tend to shift SW by $\sim$ 20 km, and the second is that the earthquake depths increase up to 40 km. These results are confirmed in all our locations, independent of the velocity model, showing that these changes are directly related to the use of OBS data. Estimated focal depths are deeper than those estimated by the IGN, and depth uncertainties are always reduced (Fig. 9a, b), even when assuming a 1D model. Note that the IGN could not resolve the depth for $\sim$ 50% of these earthquakes. In region B, the depth estimations of earthquake relocations have smaller uncertainties than those in region A and region C for any kind of relocation (Fig. 9). In region C, there are no appreciable differences either in the epicenter location or in the depth error for any kind of location. However, there is a slight improvement in the depth error estimation if the location is done using 3D models. This result is explained by the better fit of the 3D model in southern Iberia, especially in the oceanic crust, than the 1D model 1D-Iberia. Concerning the depth uncertainties in this region, they are of the same order for all locations (Fig. 9), independent of the 1D or 3D velocity model used.

The origin times show smaller differences among all locations: the RMS values range from 0.3 to 0.6 s, except for a few events (T11, T15, T16, and T30, supplementary material) where they can be as high as 1.2 s. In general, the RMSs obtained in this study are lower (Supplementary Material Tables 2, 3 and 4) than those reported by the IGN Bulletin (Supplementary Material, Table 5).

Overall, model 3D-1 determines locations with slightly smaller focal parameter uncertainties than model 3D-2 in all regions, and the influence of the OBS array is remarkable in region B (Fig. 9). The fact that there are no further improvements for any type of location in region C supports the idea that the influence of the OBS array is the most decisive factor.

Despite the small number of polarities used to estimate our fault-plane solutions (from 11 to 19), we observe good agreement between our solutions and the focal mechanisms of large earthquakes. The regional stress pattern derived from the focal mechanisms of large earthquakes shows horizontal compression in the NW–SE to NNW–SSE directions in the vicinity of CSV and the GC, a change to strike-slip motion with horizontal extension close to the Strait of Gibraltar, and only horizontal extension in the Betics east of Gibraltar (Bezzeghoud and Buforn 1999; Buforn et al. 2004). Events T28 and T34 have fault-plane solutions similar to that of the 2009 earthquake ($M_w = 5.5$). The solution for event T18 is very close to that of the 1964 earthquake ($M_w = 6.4$). It is important to note that the observed change in the focal mechanisms from thrust solutions in the CSV region and the Gulf of Cadiz to strike-slip approaching the Strait of Gibraltar (SG) and northern Morocco (Bezzeghoud and Buforn 1999; Buforn et al. 2004) is confirmed here (events T1 and T11). Finally, event T17, located in southern Spain, has a mechanism of normal oblique faulting with a strike-slip component. From Fig. 7, we can observe how the OBS array helps to constrain focal mechanisms for events T1, T11, T28, T33 and T34.

Previous works have revealed the occurrence of active seismicity (low magnitude) in region B (Geissler et al. 2010; Grevemeyer et al. 2016; Silva et al. 2017) and debated the thickness of the seismogenic layer and the spread of the epicenters in two clusters (southwest of CSV and east of GB). Geissler et al. (2010) suggested that seismicity is spread over the crust and upper mantle up to 60 km depth, although most seismicity is confined to the upper mantle between 40 and 55 km depth, which is at the limit of brittle deformation at the maximum depth of faulting ($T < 600$ °C) in the old oceanic lithosphere (McKenzie et al. 2005; Craig et al. 2014). Conversely, the results by Grevemeyer et al. (2016) have shifted the seismicity to shallower depths (20–40 km), which is congruent with the thermal model of Grevemeyer et al. (2009).

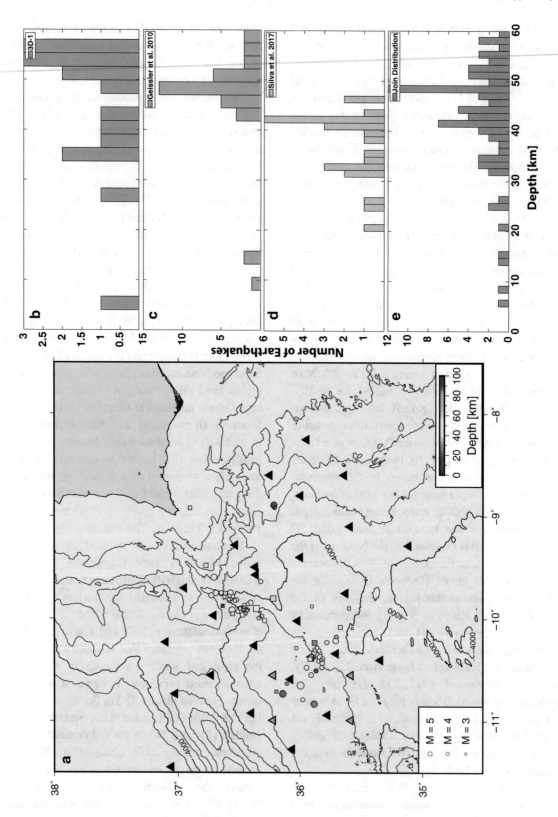

◄Figure 10
Joint hypocenter location and histogram comparison. **a** Hypocenter locations 3D-1 [squares], locations from Geissler et al. (2010) and Silva et al. (2017) [circles], OBSs of this work [green triangles], OBSs location from Geissler et al. (2010) [black triangles]. **b** Histogram with the number of earthquakes as a function of depth located in this study using 3D-1, **c** earthquakes located in Geissler et al. (2010), **d** earthquakes located exclusively in Silva et al. (2017), and **e** histogram with the joint location of 3D1, Geissler et al. (2010) and Silva et al. 2017

The results of our work in region B support the results of Geissler et al. (2010), since hypocentral locations obtained with both 1D and 3D models indicate that seismicity can be deep, reaching 51–58 km. Furthermore, region B contains the 1969 (M_s = 8.1), 2007 (M_w = 5.9) and 2009 (M_w = 5.5) earthquakes. These shocks have been studied by different authors using different methods and data (López Arroyo and Udias 1972; McKenzie 1972; Fukao 1973; Udias et al. 1976; Grimison and Chen 1986, 1988; Buforn et al. 1988a; Grandin et al. 2007, Stich et al. 2005, 2007; Custodio et al. 2012; Pro et al. 2013, Buforn et al. 2019), and their focal depths are estimated from 30 to 50 km. Geissler et al. (2010) and Silva et al. (2017) estimated focal mechanisms for earthquakes (M < 4.8) recorded on a regional marine survey carried out in the SW CSV, obtaining strike-slip and reverse dip-slip solutions. However, it is important to remark that in low-magnitude earthquakes, such as those used by Geissler et al. (2010), Silva et al. (2017) or this study (all cases with magnitudes less than 4.8), the focal mechanisms may indicate motion on small faults present in the region with orientations or motions different from the regional stress pattern derived from large earthquakes (M_w > 6.0).

Figure 10a displays the hypocenter distribution from this work (squares) together with the joint hypocenter distributions of Geissler et al. (2010, circles) and Silva et al. (2017, circles), which agree with the two clusters observed by Custódio et al. (2016), one SW of CSV and the other east of the HAP. The combination of hypocentral location results in our study and those in Geissler et al. 2010 suggests an asymmetric depth distribution (Fig. 8f) with a characteristic long tail towards shallow depths

and a clear mode at 48 km. Our data set is too sparse to discuss geological causes that may be responsible for this feature.

5. Conclusions

The use of OBSs allows us to record signals of offshore earthquakes at a closer distance and thus to locate earthquakes in regions where the azimuth gap is large and the seismicity occurs far offshore. A notable decrease in earthquake parameter location errors by up to ∼ 50% is observed in comparison with the locations from the IGN catalog in the region near the OBS array.

The use of 3D models, which better reproduce crustal complexity, is important to better constrain the hypocenter depth parameter. Furthermore, the use of a nonlinear location algorithm (NLL) helps to obtain depth estimations with sparse networks in this region.

The joint processing of OBSs data, deployed during the temporary experiment ALERTES-RIM, and land data from Spanish, Portuguese and Moroccan stations has led to the detection and more accurate location of 38 local earthquakes, one of which had previously been unknown. Most of these earthquakes are spatially confined within two clusters.

Earthquake depths show an asymmetric distribution, including a shallow cluster at ∼ 10 km and a deep cluster in the upper mantle, with depths to ∼ 55 km. These results are in agreement with those of previous studies in the same area (Geissler et al. 2010; Grevemeyer et al. 2016; Silva et al. 2017).

We obtain focal mechanisms and moment tensors (modeling land and OBS data) for seven weak earthquakes (magnitude, mbLg ≤ 4.8) and model the first-motion polarities and waveforms recorded at both OBSs and land stations; focal mechanisms and deviatoric MTs are in general agreement and show the typical transition from thrust to strike-slip motion when moving from west of CSV to the Strait of Gibraltar (Bezzeghoud and Buforn 1999; Buforn et al. 2004; Custódio et al. 2016). These results are also in agreement with the solutions for large earthquakes occurring in this region in recent decades.

In summary, the uses of sparse OBS arrays, 3D models and nonlinear location algorithms are key points for improving the geometry of seismic sources in offshore seismogenic domains.

Acknowledgements

This work was supported in part by the Spanish Ministerio de Economía, Industria y Competitividad project CGL2013-45724-C3-3-R and CGL2017-86097-R. We want to thank the Royal Spanish Navy Observatory (ROA), the Universidad Complutense de Madrid (UCM), the Instituto Geográfico Nacional (IGN, https://doi.org/10.7914/sn/es) and the Instituto Português do Mar e da Atmosfera (IPMA, https://doi.org/10.7914/sn/pm) for providing part of data, and Civiero and Grandin for providing the 3D velocity models. Figures 1, 2, 4, 8 and 9 were generated with the Generic Mapping Tools (GMT) open-source mapping toolbox (Wessel et al. 2013). We also thank the reviewers, Prof. W. Geissler and an anonymous reviewer, for their revisions that helped to improve the quality of this paper.

Publisher's Note Springer Nature remains neutral with regard to jurisdictional claims in published maps and institutional affiliations.

References

Baptista, M. A., & Miranda, J. M. (2009). Revision of the Portuguese catalog of tsunamis. *Natural Hazards and Earth System Sciences, 9*(1):25–42. http://www.nat-hazards-earth-syst-sci.net/9/25/2009/.

Bezzeghoud, M., & Buforn, E. (1999). Source parameters of the 1992 Melilla (Spain, MW = 4.8), 1994 Alhoceima (Morocco, MW = 5.8), and 1994 Mascara (Algeria, MW = 5.7) earthquakes and seismotectonic implications. *Bulletin of the Seismological Society of America, 89,* 359–372.

Bratt, S. R., & Bache, T. C. (1988). Locating events with sparse network of regional arrays. *Bulletin of the Seismological Society of America, 78,* 780–797.

Brillinger, D. R., Udias, A., & Bolt, B. A. (1980). A probability model for regional focal mechanism solutions. *Bulletin of the Seismological Society of America, 70,* 149–170.

Buforn, E., Bezzeghoud, M., Udias, A., & Pro, C. (2004). Seismic sources on the Iberia-African plate boundary and their tectonic implications. *Pure and Applied Geophysics, 161,* 623–646.

Buforn, E., López Sámchez, C., Lozano, L., Martínez Solares, J. M., Cesca, S., Oliveira, C. S., et al. (2019). Re-evaluation of seismic intensities and relocation of 1969 S. Vincent Cape seismic sequence. A comparison with the 1755 Lisbon Earthquake. *Pure Applied Geophysics.* https://doi.org/10.1007/s00024-019-02336-8.

Buforn, E., Udías, A., & Colombás, M. A. (1988a). Seismicity, source mechanism and tectonics of the Azores-Gibraltar plate boundary. *Tectonophysics, 152,* 89–118.

Buforn, E., Udias, A., & Mezcua, J. (1988b). Seismicity and focal mechanisms in south Spain. *Bulletin of the Seismological Society of America, 78,* 2008–2024.

Cabieces, R., García, A., Pazos, A., & Krüger, F. (2018). Characterization and applicability of an Ocean Bottom Seismometer Array to detect incoherent seismic signals. *EGU General Assembly Conference Abstracts* (Vol. 20, p. 9901).

Cabieces, R., Krüger. F., Garcia-Yeguas, A., Villaseñor, A., Buforn, E., Pazos, A., Olivar-Castaño, A., & Barco, J. (2020). Slowness vector estimation over large-aperture sparse arrays with the Continuous Wavelet Transform: Application to Ocean Bottom Seismometers. *Geophysical Journal International* **(Submitted Feb 2020)**.

Civiero, C., Strak, V., Custódio, S., Silveira, G., Rwlinson, N., Arroucau, P., et al. (2018). A common Deep source for upper-mantle upwelling below the Ibero-western Maghreb region from teleseismic P-wave travel-time tomography. *Earth and Planetary Science Letters, 499,* 157–172.

Craig, T. J., Copley, A., & Jackson, J. (2014). A reassessment of outer-rise seismicity and its implications for the mechanics of oceanic lithosphere. *Geophysical Journal International, 197,* 63–89. https://doi.org/10.1093/gji/ggu013.

Custodio, S., Cesca, S., & Heimman, S. (2012). Fast Kinematic Waveform Inversion and Robustness Analysis: Application to the 2007 Mw 5.9 Horseshoe Abyssal Plain Earthquake Offshore Southwest Iberia. *Bulletin of the Seismological Society of America, 102,* 361–376. https://doi.org/10.1785/0120110125.

Custódio, S., Lima, V., Vales, D., Cesca, S., & Carrilho, F. (2016). Imaging active faulting in a region of distributed deformation from the joint clustering of focal mechanisms and hypocentres: Application to the Azores–western Mediterranean region. *Tectonophysics, 676,* 70–89.

Fernandes, R. M. S., Ambrosius, B. A. C., Noomen, R., Bastos, L., Wortel, M. J. R., Spakman, W., et al. (2003). The relative motion between Africa and Eurasia as derived from ITRF2000 and GPS data. *Geophysical Research Letters.* https://doi.org/10.1029/2003gl017089.

Font, Y., Kao, H., Lallemand, S., Liu, C.-S., & Chiao, L.-Y. (2004). Hypocentre determination offshore of eastern Taiwan using the Maximum Intersection method. *Geophysical Journal International, 158,* 655–675.

Fukao, Y. (1973). Thrust faulting at a lithosphere plate boundary: The Portugal earthquake of 1969. *Earth and Planetary Science Letters, 18,* 205–216.

Geissler, W. H., Matias, L., Stich, D., Carrilho, F., Jokat, W., Monna, S., et al. (2010). Focal mechanisms for sub-crustal earthquakes in the Gulf of Cadiz from a dense OBS deployment. *Geophysical Research Letters.* https://doi.org/10.1029/2010GL044289.

Gouedard, P., Seher, T., McGuire, J. J., Collins, J. A., & van der Hilst, R. D. (2014). Correction of ocean-bottom seismometer instrumental clock errors using ambient seismic noise. *Bulletin of the Seismological Society of America, 104,* 1276–1288.

Grandin, R., Borges, J. F., Bezzeghoud, M., Caldeira, B., & Carrilho, F. (2007). Simulations of strong ground motion in SW Iberia for the 1969 February 28 (Ms = 8.0) and the 1755 November 1 (M ∼ 8.5) earthquakes—I. Velocity model. *Geophysical Journal International, 171,* 1144–1161.

Grevemeyer, I., Kaul, N., & Kopf, A. (2009). Heat flow anomalies in the Gulf of Cadiz and off Cape San Vincente, Portugal. *Marine and Petroleum Geology, 26,* 795–804. https://doi.org/10.1016/j.marpetgeo.2008.08.006.

Grevemeyer, I., Lange, D., Villinger, H., Custódio, S., & Matias, L. (2017). Seismotectonics of the Horseshoe Abyssal Plain and Gorringe Bank, eastern Atlantic Ocean: Constraints from ocean bottom seismometer data. *Journal of Geophysical Research: Solid Earth, 122,* 63–78.

Grevemeyer, I., Matias, L., & Silva, S. (2016). Mantle earthquakes beneath the South Iberia continental margin and Gulf of Cadiz—Constraints from an onshore–offshore seismological network. *Journal of Geodynamics, 99,* 39–50.

Grimison, N. L., & Chen, W. P. (1986). The Azores Gibraltar Plate Boundary: Focal mechanisms, depth of earthquakes and their tectonic implications. *Journal of Geophysical Research, 91,* 2029–2047.

Grimison, N. L., & Chen, W. P. (1988). Source mechanisms of four recent earthquakes along the Azores–Gibraltar plate boundary. *Geophysical Journal, 92,* 391–401.

Heimann, S., Isken, M., Kühn, D., Sudhaus, H., Steinberg, A., Vasyura-Bathke, H., Daout, S., Cesca, S., Dahm, T. (2018). Grond—A probabilistic earthquake source inversion framework. V. 1.0. GFZ Data Services. https://doi.org/10.5880/GFZ.2.1.2018.003.

Hudson, J. A., Pearce, R. G., & Rogers, R. M. (1989). Source type plot for inversion of the moment tensor. *Journal of Geophysical Research: Solid Earth, 94,* 765–774.

Husen, S., Kissling, E., Deichmann, N., Wiemer, S., Giardini, D., & Baer, M. (2003). Probabilistic earthquake location in complex three-dimensional velocity models: Application to Switzerland. *Journal of Geophysical Research: Solid Earth, 108*(B22), 2077–2102. https://doi.org/10.1029/2002JB001778.

Jordan, T. H., & Sverdrup, K. A. (1981). Teleseismic location techniques and their application to earthquake clusters in the south-central Pacific. *Bulletin of the Seismological Society of America, 71,* 1105–1130.

Lomax, A. (2005). Rapid estimation of rupture extent for large earthquakes: Application to the 2004, M9 Sumatra-Andaman mega-thrust. *Geophysical Research Letters.* https://doi.org/10.1029/2005gl022437.

Lomax, A. (2008). Location of the focus and tectonics of the focal region of the California earthquake of 18 April 1906. *Bulletin of the Seismological Society of America, 98,* 846–860.

Lomax, A., & Curtis, A. (2001). Fast, probabilistic earthquake location in 3D models using oct-tree importance sampling. In *Geophys. Res. Abstr* (Vol. 3, p. 955).

Lomax, A., Zollo, A., Capuano, P., & Virieux, J. (2001). Precise, absolute earthquake location under Somma-Vesuvius volcano using a new three-dimensional velocity model. *Geophysical Journal International, 146,* 313–331.

López, C. (2008). *Nuevas fórmulas de magnitud para la Península Ibérica y su entorno.* Trabajo de investigación del Máster en Geofísica y Meteorología. Departamento de Física de la Tierra,

Astronomía y Astrofísica I. Universidad Complutense de Madrid. Madrid.

López Arroyo, A., & Udias, A. (1972). Aftershock sequence and focal parameters of the February 28, 1969 earthquake of the Azores-Gibraltar fracture zone. *Bulletin of the Seismological Society of America, 62,* 699–719.

Martinez Solares, J. M. (1995). *Boletín de sismos próximos 1995* (p. 13). Madrid: IGN.

Martínez Solares, J. M., & Arroyo, A. L. (2004). The great historical 1755 earthquake. Effects and damage in Spain. *Journal of Seismology, 8,* 275–294.

Martínez-Loriente, S., Gràcia, E., Bartolomé, R., Perea, H., Klaeschen, D., Dañobeitia, J. J., et al. (2018). Morphostructure, tectono-sedimentary evolution and seismic potential of the Horseshoe Fault, SW Iberian Margin. *Basin Research, 30,* 382–400.

Martínez-Loriente, S., Gracia, E., Bartolome, R., Sallarès, V., Connors, C., Perea, H., et al. (2013). Active deformation in old oceanic lithosphere and significance for earthquake hazard: Seismic imaging of the Coral Patch Ridge area and neighboring abyssal plains (SW Iberian Margin). *Geochemistry, Geophysics, Geosystems, 14,* 2206–2231.

McKenzie, D. (1972). Active tectonics of the Mediterranean region. *Geophysical Journal International, 30,* 109–185.

McKenzie, D., Jackson, J., & Priestley, K. (2005). Thermal structure of oceanic and continental lithosphere. *Earth and Planetary Science Letters, 233*(3–4), 337–349.

Nocquet, J. M. (2012). Present-day kinematics of the Mediterranean: A comprehensive overview of GPS results. *Tectonophysics, 579,* 220–242.

Palomeras, I., Carbonell, R., Flecha, I., Simancas, F., Ayarza, P., Matas, J., et al. (2008). The nature of the lithosphere across the Variscan Orogen of SW-Iberia: dense wide-angle seismic reflection data. *Journal of Geophysical Research: Solid Earth., 114,* B02302. https://doi.org/10.1029/2007JB005050.

Podvin, P., & Lecomte, I. (1991). Finite difference computation of traveltimes in very contrasted velocity models: a massively parallel approach and its associated tools. *Geophysical Journal International, 105,* 271–284.

Pro, C., Buforn, E., Bezzeghoud, M., & Udías, A. (2013). The earthquakes of 29 July 2003, 12 February 2007, and 17 December 2009 in the region of Cape Saint Vincent (SW Iberia) and their relation with the 1755 Lisbon earthquake. *Tectonophysics, 583,* 16–27.

Sallarès, V., Gailler, A., Gutscher, M. A., Graindorge, D., Bartolomé, R., Gràcia, E., et al. (2011). Seismic evidence for the presence of Jurassic oceanic crust in the central Gulf of Cadiz (SW Iberian margin). *Earth and Planetary Science Letters, 311,* 112–123.

Sallarès, V., Martínez-Loriente, S., Prada, M., Gràcia, E., Ranero, C. R., Gutscher, M. A., et al. (2013). Seismic evidence of exhumed mantle rock basement at the Gorringe Bank and the adjacent Horseshoe and Tagus abyssal plains (SW Iberia). *Earth and Planetary Science Letters, 365,* 120–131.

Sambridge, M., & Mosegaard, K. (2002). Monte Carlo methods in geophysical inverse problems. *Reviews of Geophysics, 40,* 3-1–3-29.

Sartori, R., Torelli, L., Zitellini, N., Peis, D., & Lodolo, E. (1994). Eastern segment of the Azores-Gibraltar line (central-eastern Atlantic): An oceanic plate boundary with diffuse compressional deformation. *Geology, 22,* 555–558.

Sens-Schönfelder, C. (2008). Synchronizing seismic networks with ambient noise. *Geophysical Journal International, 174,* 966–970.

Serpelloni, E., Vannucci, G., Pondrelli, S., Argnani, A., Casula, G., Anzidei, M., et al. (2007). Kinematics of the Western Africa-Eurasia plate boundary from focal mechanisms and GPS data. *Geophysical Journal International, 169,* 1180–1200.

Shiobara, H., Nakanishi, A., Shimamura, H., Mjelde, R., Kanazawa, T., & Berg, E. W. (1997). Precise positioning of ocean bottom seismometer by using acoustic transponder and CTD. *Marine Geophysical Researches, 19,* 199–209.

Silva, S., Terrinha, P., Matias, L., Duarte, J. C., Roque, C., Ranero, C. R., et al. (2017). Micro-seismicity in the Gulf of Cadiz: Is there a link between micro-seismicity, high magnitude earthquakes and active faults? *Tectonophysics, 717,* 226–241.

Stachnik, J. C., Sheehan, A. F., Zietlow, D. W., Yang, Z., Collins, J., & Ferris, A. (2012). Determination of New Zealand ocean bottom seismometer orientation via Rayleigh-wave polarization. *Seismological Research Letters, 83,* 704–713.

Stähler, S. C., Sigloch, K., Hosseini, K., Crawford, W. C., Barruol, G., Schmidt-Aursch, M. C., et al. (2016). Preliminary performance report of the RHUM-RUM ocean bottom seismometer network around La Réunion, western Indian Ocean. *Advances in Geosciences, 41,* 43–63.

Stehly, L., Campillo, M., & Shapiro, N. M. (2007). Traveltime measurements from noise correlation: stability and detection of instrumental time-shifts. *Geophysical Journal International, 171,* 223–230.

Stich, D., de Lis Mancilla, F., Pondrelli, S., & Morales, J. (2007). Source analysis of the February 12th 2007, Mw 6.0 Horseshoe earthquake: Implications for the 1755 Lisbon earthquake. *Geophysical Research Letters.* https://doi.org/10.1029/2005gl023098.

Stich, D., Mancilla, F. D. L., & Morales, J. (2005). Crust-mantle coupling in the Gulf of Cadiz (SW-Iberia). *Geophysical Research Letters.* https://doi.org/10.1029/2005GL023098.

Stich, D., Martínez-Solares, J. M., Custódio, S., Batlló, J., Martín, R., Teves-Costa, P., & Morales, J. (2020). Seismicity of the Iberian Peninsula. In *The geology of Iberia: A geodynamic approach* (pp. 11–32). Springer, Cham.

Tarantola, A., & Valette, B. (1982). Inverse problems = quest for information. *Journal of Geophysics, 50,* 159–170.

Udias, A., Arroyo, A. L., & Mezcua, J. (1976). Seismotectonic of the Azores-Alboran region. *Tectonophysics, 31,* 259–289.

Webb, S. C. (1998). Broadband seismology and noise under the ocean. *Reviews of Geophysics, 36,* 105–142.

Wessel, P., Smith, W. H., Scharroo, R., Luis, J., Wobbe, F. (2013). Generic mapping tools: improved version released. *Eos, Transactions American Geophysical Union, 94,* 409–410.

Zhou, H. W. (1994). Rapid three-dimensional hypocentral determination using a master station method. *Journal of Geophysical Research: Solid Earth, 99*(B8), 15439–15455.

Zitellini, N., Gràcia, E., Matias, L., Terrinha, P., Abreu, M. A., Dealteriis, G., et al. (2009). The quest for the Africa-Eurasia plate boundary west of the Strait of Gibraltar. *Earth and Planetary Science Letters, 280,* 13–50.

(Received October 13, 2019, revised March 21, 2020, accepted March 25, 2020, Published online April 14, 2020)

Pure Appl. Geophys. 177 (2020), 1781–1800
© 2019 Springer Nature Switzerland AG
https://doi.org/10.1007/s00024-019-02336-8

Pure and Applied Geophysics

Re-evaluation of Seismic Intensities and Relocation of 1969 Saint Vincent Cape Seismic Sequence: A Comparison with the 1755 Lisbon Earthquake

E. Buforn,[1] C. López-Sánchez,[1] L. Lozano,[2] J. M. Martínez-Solares,[2] S. Cesca,[3] C. S. Oliveira,[4] and A. Udías[1]

Abstract—Seismic intensity for the February 28, 1969 (Mw = 7.8) earthquake have been re-evaluated using original documents in local archives, such as, contemporary newspapers, council minutes, monographic studies, among other sources for Spain, Portugal and Morocco and answers to macroseismic questionnaires for Morocco. This information is used to plot a new intensity map for the whole region affected by the earthquake: Portugal, Spain and Morocco. The intensity values vary from VIII to IX in the E-W coast of Algarve, southern Portugal, to II–III. Furthermore, we have relocated the hypocentres for main shock and 24 aftershocks using a new 3D crustal velocity model for the Gulf of Cadiz region and a non-linear probabilistic location methodology, most of them previously lacking a depth estimate. The new locations show an E-W distribution of epicenters, with focus located in the uppermost mantle, most of them with depths between 30 and 50 km. No earthquakes have been located at depths shallower than 30 km. A comparison between peak ground accelerations (PGAs) estimated from the observed intensities for the 1969 and the Lisbon 1755 earthquakes, and synthetic PGA values, generated assuming two different scenarios (using the 1969 and 2009 earthquakes) for the 1755 Lisbon event, shows that the observed damage produced by the 1755 earthquake may be better explained assuming a reverse dip-slip mechanism oriented in NE-SW direction, similar to that of the 2009 earthquake, rather than assuming focal mechanism similar to that of the 1969 earthquake.

Key words: 1969 earthquake, 1755 Lisbon earthquake, intensity re-evaluation, hypocentral relocation.

Electronic supplementary material The online version of this article (https://doi.org/10.1007/s00024-019-02336-8) contains supplementary material, which is available to authorized users.

[1] Dto de Física de la Tierra y Astrofísica, Universidad Complutense, 28040 Madrid, Spain. E-mail: caroll04@ucm.es
[2] Instituto Geográfico Nacional, Madrid, Spain.
[3] GFZ German Research Centre for Geosciences, 14467 Potsdam, Germany.
[4] Instituto Superior Tecnico/CEris, Universidade de Lisboa, Lisbon, Portugal.

1. Introduction

The 1969, February 28 $M_w = 7.8$ earthquake is the largest shock occurred during the instrumental period (1920—present) offshore SW of the Iberian Peninsula, at the plate boundary between Eurasia and Africa and with epicenter close to that assumed for the 1755 Lisbon earthquake. The 1969 event generated a tsunami (about 1 m high), much smaller than that of 1755, reaching SW Iberia (Portugal and Spain) and the Atlantic coast of Morocco. The 1969 earthquake was felt at great part of Iberia and NW Morocco, producing the largest damage along the Atlantic coast of Portugal and Spain. There are several intensity maps for this event, most corresponding only to the intensity values for each country: Portugal, Spain or Morocco. Only a few of them show the seismic intensity for the whole region. In this study, we have carried out a revaluation of the seismic intensities for the 1969 main shock for the whole region, that is Portugal, Spain and Morocco, using original contemporary sources and the EMS-98 scale.

The 1969 earthquake was followed by a seismic sequence lasting several months. However, the epicentral locations of these aftershocks are very poorly, determined due to the low number of seismic stations at the region in that time, all very distant from the epicentral area, the poor azimuthal coverage and the use of 1D structure models for the hypocentral location. Due to these limitations, it was not possible to estimate reliable focal depths for most of these aftershocks. Over the last years, new algorithms using 3D velocity models have been developed for hypocentral determination, which allows improving the location accuracy, especially, with respect the determination of the focal depth. In the second part of this study, we discuss the relocation of a subset of the largest aftershocks, using a new accurate 3D crustal velocity distribution for the Gulf of Cadiz region (Lozano et al. 2019 under revision) and a non-linear probabilistic location method (NonLinLoc, Lomax et al. 2000). Data used correspond to all the available regional phase data recorded at Spanish, Portuguese and Moroccan stations. We have used the original arrival times, from the Instituto Geográfico Nacional (IGN) Seismic Catalogue, seismic stations bulletins from 1969 to 1970 and included a number of new available phase data information. Following several reliability criteria (a minimum number of stations and of P and S waves arrivals), we have selected 25 earthquakes occurred in the period between February 28 and December 31, 1969. Although the location of these earthquakes is highly influenced by the poor network configuration (all stations are located onshore at epicentral distances larger than 300 km and there is a large azimuthal gap of 240° of stations coverture), the use of a probabilistic approximation and a more realistic 3D crust and upper mantle model provides not only better maximum likelihood hypocentre, but also more accurate focal depths, not available by previous studies.

The 1969 main shock occurred on February 28, at 02 h 40 m 32 s with epicentre located SW of Saint Vincent Cape (35.9850°N, 10.8133°W) and at 20 km depth, (http://www.ign.es/web/ign/portal/sis-catalogo-terremotos). It is located at the plate boundary between Eurasia and Africa, very close to the point where the transition begins from an oceanic and well defined plate boundary to the west to a continental and more diffuse limit to the east (Fig. 1a, b). As a consequence of this transition, the seismicity is more diffuse from 12°W to the east, with several alignment of epicenters (Fig. 1b), corresponding the plate boundary to a wider deformation zone (Buforn et al. 1995). From 12°W to 9°W, we observe several alignments of epicentres that may be correlated with the main geological structures (Martínez-Loriente et al. 2013). A NE-SW distribution of epicentres may be correlated to the Gorringe Bank Fault (GBF), another alignment of epicentres follows a WNW–SES direction and may be associated to the Horseshoe Fault (HF) where the 1969, the 2003 ($M_w = 5.3$) and the 2007 ($M_w = 5.9$)

earthquakes are located (Fig. 1b). There is a third line of epicentres, along the submarine Saint Vincent Canyon Fault (SVCF) that starts around 36°N, 10°W and runs in a NE-SW direction to the western part of Saint Vincent Cape (SVC) where is located the 2009 earthquake ($M_w = 5.5$). From 9°W to 7°W the epicentres follow an E-W direction over a wider zone (near 100 km width) following the Guadalquivir Bank (GB), with the seismic activity decreasing as it approach the Spanish coast. In this zone is located the $M_w = 6.5$, 1964 earthquake (Udías and López Arroyo 1970). A NW-SE minor alignment may be observed from 8°W to 6°W, following the Gulf of Cadiz-Morocco Fault (CGMF) which reaches the interior of Morocco; the largest earthquake located in this zone is the 1960 event ($M_b = 6.1$). The location of the large 1755 Lisbon earthquake is still a debatable question and several macroseismic epicentres have been proposed based on the distribution of damage, tsunami propagation and other factors (Machado 1966; Moreira 1985; Martínez Solares et al. 1979; Baptista et al. 1998; Zitellini et al. 1999; Vilanova et al. 2003; Martínez Solares and López Arroyo 2004; Grandin et al. 2007; Pro et al. 2013). We use in this study the epicentre proposed by Martínez Solares and López Arroyo (2004), located in proximity to the 2009 event. This epicentre is based on the results obtained from modelling of the tsunami waves generated by the earthquake (Baptista et al. 1998). Further, the intensities distribution for the 2009 earthquake show high values along the south of Portugal (the Algarve N–S coast) and in the Lisbon region, similar to those of the 1755 earthquake (Carranza 2016).

The 1969 shock has been the object of different studies with similar results on focal mechanism: reverse faulting with both nodal planes oriented in approximately E–W direction and horizontal pressure axis oriented in NNW-SSE direction. The focal depth varies from 16 to 33 km (López-Arroyo and Udías 1972; McKenzie 1972; Fukao 1973; Udías et al. 1976; Grimison and Chen 1986, 1988; Buforn et al. 1988; Grandin et al. 2007). The 1964 earthquake, 1969 largest aftershock (1969A in Fig. 1b), 2003 and 2007 events have all similar focal mechanisms. However, the 2009 earthquake, which is located at NE and with similar epicenter of those proposed by

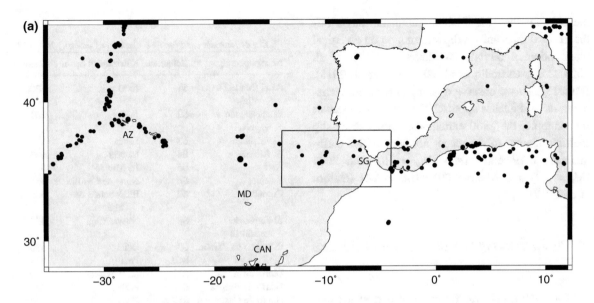

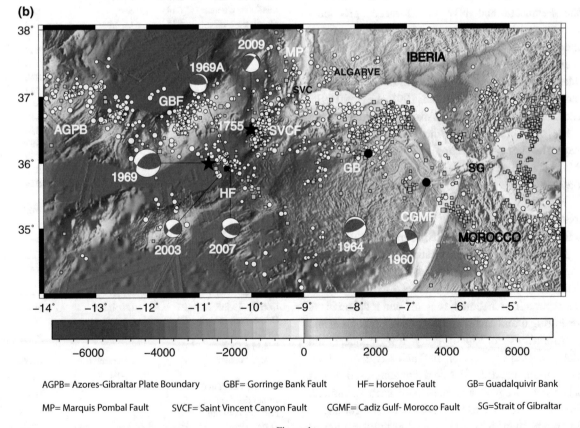

AGPB= Azores-Gibraltar Plate Boundary   GBF= Gorringe Bank Fault   HF= Horsehoe Fault   GB= Guadalquivir Bank

MP= Marquis Pombal Fault   SVCF= Saint Vincent Canyon Fault   CGMF= Cadiz Gulf- Morocco Fault   SG=Strait of Gibraltar

Figure 1

a Seismicity along the plate boundary between Eurasia and Africa from the Azores Triple Point to Tunisia for period 1930–2019, M ≥ 5.0. The square corresponds to the region plotted in Fig. 1b. *AZ* Azores, *MD* Madeira, *CAN* Canary Islands, *SG* Strait of Gibraltar. **b** Distribution of epicenters for the period 2000–2019 (M ≥ 3.0) (https://www.ign.es/web/ign/portal/sis-catalogo-terremotos) (white circles h < 40 km, grey squares, h > 40 km) and focal mechanism of largest shocks for this area (Pro et al. 2013). Stars correspond to the 1755 (Martínez Solares and López Arroyo 2004) and 1969 epicenters (https://www.ign.es/web/ign/portal/sis-catalogo-terremotos)

the IGN for the 1755 Lisbon earthquake has a different focal mechanism: dip-slip motion on a vertical plane oriented on NE-SW direction (Stich et al. 2005, 2007; Custodio et al. 2012; Pro et al. 2013). From Fig. 1b we observe a change on focal mechanisms as earthquakes approach the Strait of Gibraltar: for example, the 1960 earthquake shows strike-slip faulting, similar to the recent northern Africa earthquakes: Al-Hoceima 1994 ($M_w = 5.8$), and 2004 ($M_w = 6.2$) and Alboran 2016 ($M_w = 6.4$) (Buforn et al. 2017).

2. Re-evaluation of Intensities for the 1969 Main Shock

The 1969 shock was felt in great part of the Iberian Peninsula and NW of Morocco, producing damage and losses of human life. There are several intensity maps for this earthquake, but most show the intensities for Portugal, Spain or Morocco separately (Martínez Solares et al. 1979; Moreira 1984; Paula and Oliveira 1996; Cherkaoui 1991; Levret 1991). The intensity map by López-Arroyo and Udías (1972) is the only one covering the whole region. Other maps for the whole region are based on this one (Martínez Solares et al. 1979; Mézcua 1982) or put together joining several maps (Grandin et al. 2007).

For this reason, we have carried out a search of contemporary documents in local and national archives, archives of city halls, churches, etc. in order to find detailed information about the damage produced by this shock. We have used newspapers from Spain (54), Portugal (10), Morocco (3) and France (1), reports from national, regional and local archives for Portugal and Spain and answers to macroseismic questionnaires for Spain (very low number) and Morocco (larger number). In Tables 1, 2 and 3 we have summarized the list of documents used in this paper and assigned a reference to each document. From these documents we have obtained information for more than 640 sites on the three countries including data from as far as Azores, Madeira and Canary Islands. Based on this data we have re-evaluated the seismic intensities at a large number of places using the EMS-98 scale (Table S1). In addition, we have complemented the intensity values with

Table 1

List of journals used for re-evaluation of intensity for 1969

Newspapers Spain	Reference	Newspapers Spain	Reference
ABC (Madrid and Sevilla)	S1	Ideal	S30
El adelantado de Segovia	S2	Jaén	S31
El Adelanto	S3	Lanza	S32
El Alcázar	S4	Madrid	S33
Área	S5	Le Monde	S34
Baleares	S6	Norte de Castilla	S35
Córdoba	S7	El Noticiero de Cartagena	S36
El correo de Andalucía	S8	Nueva Rioja	S37
El correo de Zamora	S9	Odiel	S38
El día	S10	Proa	S39
Diario de Ávila	S11	La Provincia	S40
Diario de Burgos	S12	Pueblo	S41
Diario de Cádiz	S13	Soria	S42
Diario de Cuenca	S14	Sur	S43
Diario de Las Palmas	S15	La tarde	S44
Diario de León	S16	El telegrama de Melilla	S45
Diario de Navarra	S17	La Vanguardia	S46
Diario Montañés	S18	La verdad	S47
Diario Palentino	S19	La voz de Albacete	S48
Diario popular	S20	La voz de Almería	S49
Diario regional	S21	La voz de Asturias	S50
Diario vasco	S22	La voz de Castilla	S51
Extremadura	S23	La voz de Galicia	S52
El faro de Ceuta	S24	La voz del Sur	S53
Faro de Vigo	S25	Ya	S54
Gaceta del Norte	S26		
La higuerita	S27	Macroseismic enqueries	MES
Heraldo de Aragón	S28		
Hoy	S29		

Newspapers Portugal	Reference	Newspapers Morocco	Reference
Correio do Sul	P1	L'opinion	M1
Democracia do sul	P2	Le petit Marocain	M2
Diario de Lisboa	P3	La vigie marocaine	M3
Diario do sul	P4		
Diario Popular	P5	Macroseismic enqueries	MEM
Folha do domingo	P6		
Jornal do Algarve	P7		
Noticias d'Evora	P8		
O Algarve	P9		
República	P10		
Povo Algarve	P11		

Table 2

Local archives used for re-evaluation of intensity for 1969

ADPH: Archivo de la Diputación Provincial de Huelva
AMH: Archivo Municipal de Huelva
AMIC: Archivo Municipal de Isla Cristina (Huelva)
BOOH: Boletín Oficial del Obispado de Huelva
BPPH: Biblioteca Pública Provincial de Huelva
HBPPC: Hemeroteca de la Biblioteca Pública Provincial de Cádiz
HMJF: Hemeroteca Municipal de Jerez de la Frontera (Cádiz)
HN: Hemeroteca Nacional

Table 3

Local references used for re-evaluation of intensity for 1969

Alvarez Checa, J., De la Villa Márquez, L. and Mojarro Bayo, A.
    M. (2002) *Guía de arquitectura de Huelva*. Colegio Oficial de
    Arquitectos de Huelva, 360 pp
Carrasco Terriza, M. J., Gónzalez, Gómez, J. M., Oliver, A.,
    Pleguezuelo, A., Sánchez Sánchez, J. M. (2006). *Guía artística
    de Huelva y su provincial*. Fundación José Manuel Lara.
    Diputación Provincialde Huelva, 632 pp
Díaz Hierro, J. (1975). *Historia de la Merced de Huelva, hoy
    Catedral de su Diócesis*. Editorial Huelva
Fernández Jurado, J. (1986). *Huelva y su provincia*. Ediciones
    Tartessos S.L. Volumen I, 325 pp
Junta de Andalucía (1991). *La Merced. Cuatro siglos de historia*.
    Editorial Artes Gráficas Girón. 114 pp
Lara Ródenas, M. J. (2005). *Biografía de una iglesia: la parroquia
    de la Concepción de Huelva*. Colegio Oficial de Arquitectos de
    Huelva, 139 pp
Rodríguez López, J. (1991). *Isla Cristina en La Higuerita:
    recopilación de noticias, con comentarios, publicadas en "La
    Higuerita de tiempo pretérito y recuerdos"*. Isla Cristina, 144 pp
Sugrañes Gómez, E. J. (1998). *La Milagrosa y las Hijas de la
    Caridad en Huelva*. Editorial Jiménez S.L., 187 pp

those obtained by the Instituto Portugues do Mar e Atmósfera (IPMA) for other sites not included in our documents (Batlló et al. 2012), these values are reported in Table S1 with an asterisk. The total number of sites with intensity values is 746.

3. Largest Intensities

A detailed description of damages corresponding to sites with largest intensity values is given in the *Supplementary Material*, here we summarized this information. The maximum intensity value has been assigned to Fonte dos Luzeiros, a small town in Algarve (South Portugal), which was almost totally destroyed by the earthquake (of the 16 houses only one was not destroyed: P1, P3, P5, P7, P6, P9, P11, references from Table 1) (Fig. 2a). We have assigned to this town an intensity value VIII–IX (EMS-98) (Fig. 3a, b).

There is a second group of sites in south Portugal, with maximum intensity VIII: Aljezur, Barao de S. Joao, Barao de S. Miguel, Benisafrim the largest town seriously affected with about 1700 inhabitants at that time), Cacela, Fuseta and Vila do Bispo. A third group of towns have been evaluated with maximum intensity VII–VIII. They are Castro-Marim, Lagos, Sagres and Tavira, a town with important historical buildings, which were seriously affected by the shock.

We have assigned maximum intensity VII to Estoi, Faro, Loulé and Portimao (Portugal), Beas and Isla Cristina (Fig. 2b) (SW Spain). Sites with assigned intensity VI–VII include Huelva, Ayamonte, Palos de Moguer in SW Spain and in SW Portugal the monastery of Batalha, Santa Quiteria Church (Meca), and Setúbal.

There are many cities and towns with intensity VI. We summarized the damage in large cities in Portugal (Lisbon), Spain (Seville) and Morocco (Rabay and Salé). In Lisbon 58–60 people were injured, cracks in many houses, Hospital of S. José, Church of Luz (large crack and risk of collapse), Paços do Concelho, school of Bellas Artes, fall of chimneys, statues and roofs. In Seville 3 people died for cardiac attacks and damage was reported in the Cathedral (Giralda tower), the *Torre del Oro* (Tower of Gold, Fig. 2c), the pillars of the dome and the Alcazar (cracks), as well as the partial collapse of some houses, walls, etc. In Morocco the largest intensity values correspond to Rabat and Salé (at 5 km from Rabat), 5 people died, 4 were injured and 6 old houses collapsed.

The total number of casualties due to the 1969 shock was 25 deaths, with more than 80 injured. In Spain 4 people died (3 in Seville and 1 in Badajoz, due to heart attacks and about 10 were injured. In Portugal 13 people died (11 directly due to the earthquake and 2 due to consequence of heavy injures suffered) and 58 injured. In Morocco 8 people died (3 in Safi, 2 in Salé, 2 in Chichaoua, 1 in Imi n'Tanoute) and more than 12 were injured. The high number of

Figure 2
Damage of the 1969 earthquake in Fonte Luzeiros, Portugal (**a**). Isla Cristina, Spain (**b**) and Seville, Spain (**c**)

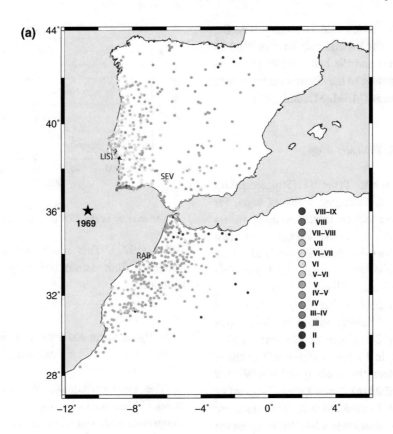

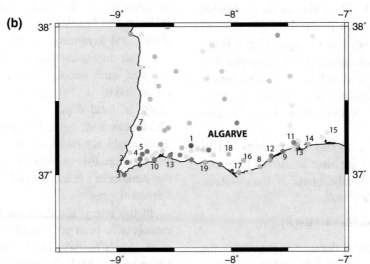

1.- Fonte dos Luzeiros   2.- Vila do Bispo   3.- Sagres   4.- B. S. Joao   5.- Bensafrim

6.- B. S. Miguel   7.- Aljezur   8.- Fuseta   9.- Cacela   10.- Lagos   11.- Castro Marim

12.- Tavira   13.- Lagos   14.- Isla Cristina   15.- Beas   16.- Estoi   17.- Faro

18.- Loulé   19.- Portimao

Figure 3
a New map of EMS-98 intensities for the 1969 earthquake. *LIS* Lisbon, *SEV* Seville, *RAB* Rabat. **b** Detail of new intensities on the SW Iberia

victims in Morocco, took place in areas where the maximum intensity was VI, and may be explained by the poor construction and the bad state of preservation of houses, according to the documents the deaths occurred in old houses (MEM, M1, M2, M3).

4. Intensity Map

In Fig. 3a and b and Table S1 (Supplementary Material) we show the assigned intensity value for each site. We observe that the largest intensity values (VII–VIII to VIII–IX) are in the southernmost coast of the Algarve region (Southern Portugal) along an E-W direction parallel to the coast located at epicentral distances between 200 and 320 km. However, at similar epicentral distance, but along the N-S coast of the Algarve, the intensity values are lower (for example at Faro or Sines both at 280 km). Further north, for example, in Lisbon (located at 340 km at NNE of the epicenter), the intensity value is VI, but in Castro-Marim (330 km), Santa Luzia (310 km) or Tavira (309 km), all located along the E-W southern coast the maximum intensity is VII–VIII. In Spain the higher intensities (VI–VII to VII) are in the Huelva region at the border with Portugal: in Isla Cristina (340 km, intensity VII), Huelva city (373 km, intensity VI–VII) and Ayamonte (334 km, intensity VI–VII),. In Morocco the distribution of intensities is very uniform; the maximum values are (VI) in Rabat-Salé (450 km) similar to those of Seville at the same epicentral distance.

In Fig. 4 we have plotted the intensity versus distance together with the error bands We have carried out a logarithmic adjustment of these values, obtaining the following relation:

$$I = 30.72 - 4.15 ln(R)$$

where R is the epicentral distance in km. Using the values plotted in Fig. 4, and assuming a circular distribution in the intensity map with radius the average distance, we have applied the methodology developed by Johnston (1996) to obtain the moment magnitude M_w. If we use the area defined by intensity III, we obtain $M_w = 7.4$, clearly lower than $M_w = 7.8$ estimated by Fukao (1973) and Grimison and Chen (1988) and $M_s = 7.8$ estimated by López-Arroyo and

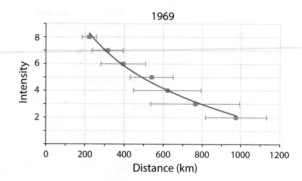

Figure 4
Attenuation of the 1969 new intensities versus distance

Udías (1972). This result shows that Mw values obtained from isoseismal maps may be underestimated.

5. Hypocentral Relocation of the 1969 Seismic Sequence

The 1969 earthquake was followed by a long series of aftershocks lasting over a year. The Instituto Geográfico Nacional (IGN) estimated the hypocentral coordinates for the main event and aftershocks (Mézcua and Martínez Solares 1983). However, the distant and asymmetric distribution of seismic stations (all analogical instruments onshore), and the complex earth structure in this region, cause large uncertainties in hypocentral determination, especially, in focal depth. In fact, for more than 90% aftershocks with epicentral determination the focal depth could not be constrained (https://www.ign.es/web/ign/portal/sis-catalogo-terremotos) due to the low number of available P and S phases and the large azimuthal gaps.

In the last years, different 2D and 3D velocity models have been proposed for this region (Grandin et al. 2007; Sallarès et al. 2011, 2013; Martínez-Loriente et al. 2013; Civiero et al. 2018; Lozano et al. 2019 under revision). The development of non linear methods for hypocentral determination, such as the NonLinLoc (NLL) software (Lomax et al. 2000; http://alomax.free.fr/nlloc/), and the use of 3D models, are powerful tools to improve the hypocentral parameter. This method performs a non-linear probabilistic search of an earthquake hypocenter, allowing travel-time calculation within a 3-D grid. The main advantage of this non-linear location is that it represents a complete, probabilistic solution of the location problem and calculates the posterior probability density function (PDF), which provides more reliable location uncertainties than classical linearized inversions (Tarantola and Valette 1982). Therefore, this method allows identifying multiple optimal solutions within irregular confidence volumes.

We have used the NLL software and an extended version of the 3D P-wave velocity model (Lozano et al. 2019 under revision) to relocate the 1969 main shock and aftershocks. This model was interpolated from a velocity-depth database retrieved from the most significant published 2-D seismic velocity models from active seismic profiles surveys carried out in the SW Iberia margin. We have extended the 3-D model propose by Lozano et al. (2019, under revision) to the rest of the Iberian Peninsula using the IGN 1-D continental-crust model (Mézcua and Martínez Solares 1983) and a 1-D homogeneous layered model for the Alboran domain based on recent local tomography studies (Stich et al. 2003; Grevemeyer et al. 2015). The new 3D P-wave velocity model (Fig. 5) covers the region from latitude 33°N to 44°N, and longitude 15°W to 4°E and down to 80 km depth. For upper-mantle depths, we have considered a laterally heterogeneous velocity distribution using three different 1-D layered models: a constant P-wave velocity of 8.15 km/s for the oceanic domain (Laske et al. 2013; Custodio et al. 2015), a local velocity model modified from Silva et al. 2017 for the Gulf of Cadiz region, and the IASP-91 global model (Kennett and Engdahl 1991) for the Iberian Peninsula which assumes a smooth velocity gradient from 8.0 km/s at the Moho.

For the relocation with NLL we generated a regional $1761 \times 1241 \times 80$ km³ grid volume with 1 km spacing interpolated using a kriging algorithm from the original 3D velocity model and assumed a "2flat-earth" approach. For the inversion, the depth search was set free to oscillate in the whole depth range and we assumed a constant average v_p/v_s ratio of 1.74, obtained using the Wadati diagram with a set of representative M > 3 earthquakes registered in the region for the period 2007–2018.

We have relocated the main shock and a selection of 24 aftershocks from the IGN catalogue occurred in the period from 28 February to 31 December 1969, recorded at the regional seismological stations from Spanish, Portuguese and Moroccan networks (Fig. 6a). We only considered those earthquakes recorded at least by 8 seismic stations and with a minimum of 8 P phases and 4 S phases. For the main shock we used 12 P-wave readings, since no S-wave were available, and for the largest aftershock 10 P-wave and 2 S-wave readings. Data corresponds to the IGN original bulletins, original readings (taken from seismograms) of P and S-waves by López-Arroyo and Udías (1972) and some additional data. During the relocation process, phases with residuals larger than 2.5 s were considered outliers and removed. In Table 4 we show the relocated hypocenters and error parameters (RMS of travel time residuals, depth errors and the semi-axes of the 90% confidence 2-D error ellipsoid). Maximum likelihood hypocenters as well as the probability density functions (PDF) are plotted in Fig. 6b. The location PDFs are large and highly irregular, mainly due to the fact that all earthquakes are located outside the network, far away from land stations (distance to the closest station over 300 km), and with significant azimuthal gaps (over 240°) (Lomax et al. 2000, 2009). Results must be taken with caution, since they may be strongly affected by the poor station distribution, the scarce number of phases used for location, and the use of a "flat-earth" approximation, resulting in large depth uncertainties.

From Fig. 6b, where we have also plotted the epicenters taken from the IGN catalogue, we observe that now the relocated main shock (the red star) is closer to the coast (black star is the IGN catalogue epicenter) and deeper (37 km versus 20 km). Fukao (1973) obtained focal depths of 16 km or 33 km for the main shock, and of 40–45 km for aftershocks. Grimison and Chen (1988) estimated a depth of 31 km for the main shock and 50 km for the largest aftershock (versus 37 km in our study). Other reported focal depths for main shock, varies from 10 km (US Geological Survey Catalogue), 19 km (ISC Catalogue) or 21 km (Centenial Catalogue, Engdahl et al. 1998; and Villaseñor and Engdahl 2005). Most aftershocks relocated in this study (red

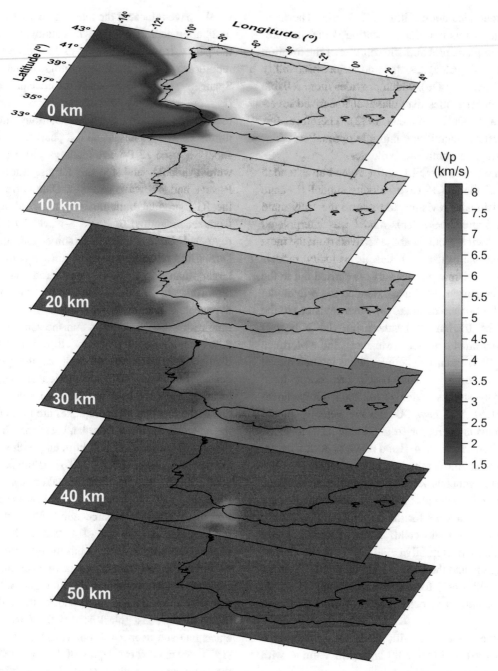

Figure 5
3-D velocity model used on this study

circles) are in a depth range between 30 and 40 km (19 events), with only three earthquakes deeper than 55 km (Table 4). No earthquakes have been located at depths shallower than 30 km. In the IGN catalogue (black crosses in Fig. 6b), depth parameter was not constrained, except for one aftershock. This range of focal depths is in good agreement with results obtained for other earthquakes in this region. For example, the 2007 (M_w = 6.0) and 2009 (M_w = 5.5) earthquakes, located in this region, have focal depth of 30–40 km

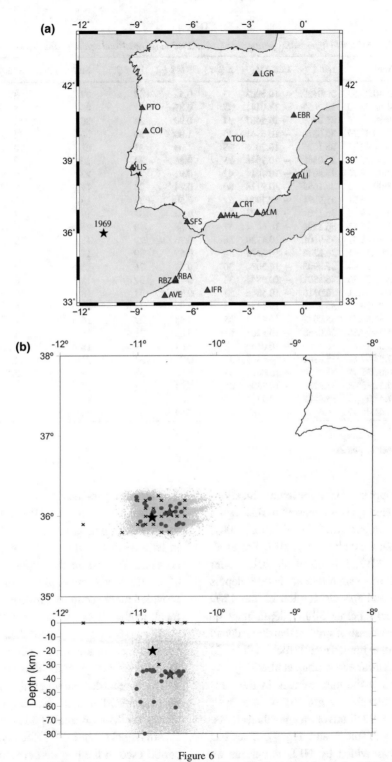

Figure 6

a Seismic stations in the Iberian Peninsula which recorded the 29 February 1969 earthquake and its aftershocks. **b** Relocation of 1969 aftershock series using NLL algorithm and a 3D structure model (Lozano et al. 2019, under revision). At top, distribution of epicenters for the 1969 main shock (black star previous solution, red star this study). Black crosses are the previous aftershocks and red circles relocated epicenters. At bottom, vertical cross section of hypocenters (symbols as top). The grey area shows the probability density functions

Table 4

Relocated hypocentral parameters for the 1969 main shock and selected aftershocks

Date	Mag	Time	Lat. (°)	Lon. (°)	h (km)	RMS (s)	Errz (km)	Smajax (km)	Sminax (km)	Az(°)	Nphs
1969/02/28	7.8[a]	02:40:34.97	36.0429	− 10.5923	37	0.52	37	37	8	19	12
1969/02/28	5.7	04:25:33.70	36.2054	− 11.0141	37	0.87	35	44	18	16	12
1969/02/28	4.6	09:59:51.57	36.0548	− 10.8087	34	0.90	34	26	13	13	13
1969/02/28	4.2	15:20:42.79	36.0310	− 10.6541	35	1.00	30	30	12	16	16
1969/02/28	3.6	16:21:02.21	36.1023	− 10.7006	38	1.03	31	28	10	11	15
1969/02/28	4.2	18:24:38.79	36.0548	− 10.7948	35	0.86	32	26	9	10	13
1969/03/01	4.1	22:07:53.72	35.9042	− 10.7014	42	0.91	33	22	9	9	17
1969/03/02	4.4	18:00:59.73	36.1658	− 10.9138	80	0.74	34	17	7	8	23
1969/03/05	4.7	02:57:36.21	35.9874	− 10.7973	57	0.50	33	18	5	14	17
1969/03/06	4.8	19:23:44.00	36.1301	− 10.8904	34	0.62	32	17	7	14	18
1969/03/07	4.3	21:31:16.75	36.1578	− 10.4674	37	0.68	30	26	6	9	16
1969/03/08	4.2	03:36:00.47	35.9161	− 10.4730	38	0.76	29	15	8	18	18
1969/03/09	4.5	13:08:16.05	36.2212	− 10.8479	34	0.88	30	17	7	8	24
1969/03/10	3.9	09:56:50.44	35.8903	− 10.5026	36	0.73	26	16	5	13	25
1969/03/12	3.7	03:11:35.26	36.1301	− 10.5840	36	0.72	32	18	8	4	16
1969/03/18	4.2	04:17:37.11	36.0310	− 10.5846	37	0.94	33	18	10	17	17
1969/03/18	4.0	06:00:36.32	35.9042	− 10.4236	35	0.80	28	24	6	12	18
1969/04/10	4.2	16:52:06.58	35.9042	− 10.4792	35	0.68	29	16	7	10	16
1969/05/05	5.5	05:34:24.85	36.0568	− 10.5201	61	0.83	31	15	6.5	13	24
1969/05/10	4.3	13:31:14.82	36.2450	− 10.0169	48	0.64	33	16	7.0	7	20
1969/08/03	4.1	02:53:07.55	35.8804	− 10.6443	36	0.75	34	23	12.5	− 6	16
1969/08/08	4.2	20:02:35.75	35.9359	− 10.9675	57	0.96	33	22	12.4	7	16
1969/10/18	4.2	05:31:45.68	36.0231	− 10.8205	35	0.64	32	24	10.8	− 6	14
1969/11/05	4.6	07:47:34.73	35.8883	− 10.9364	35	0.85	29	19	8.4	19	19
1969/12/24	5.1	05:04:46.79	36.0548	− 10.4469	37	0.69	32	20	10.0	3	13

[a]Moment magnitude M_w

Az azimuth, *Nphs* number of phases

and the same happens for hypocenter locations obtained for other earthquakes occurred in this region in the NEAREST project (Stich et al. 2005, 2007; Geissler et al. 2010; Custodio et al. 2012; Pro et al. 2013; Silva et al. 2017; Lozano et al. 2019 under revision). The lack of seismic foci at shallow depths (less than 30 km) and the distribution of the 1969 sequence with shocks below 30 km depth may be explained due to the release of stress at that depth due to the existence there of exhumed mantle rocks, with rigid and cool material (Martínez-Loriente et al. 2013).

In Fig. 7 we show the uncertainties in the relocations. In general, the origin times are well estimated, with most RMS travel time residuals in the range of 0.6–1.0 s (Table 4 and Fig. 7). The 2-D confidence ellipse provided by NLL represents an accurate image of the uncertainties. Epicentral errors (Smajax and Sminax) vary from 14 to 30 km and 5 to 1 km respectively. From Table 4 we can also observe

that most ellipse-error azimuths are oriented almost N–S (between − 6° and 19°). So, we can conclude that the epicenters are better located in longitude than in latitude due to the poor geometry of the stations network. The use in the hypocentral determinations of a non-linear probabilistic location method has provided more accurate solutions, in spite of the large focal depths errors (of the order of 25–35 km).

In Fig. 8 we have plotted the aftershock sequence versus time, starting on 28 February. Data are taken from López-Arroyo and Udías (1972) and correspond to events recorded, although the epicenters could not be determined due to the low number of seismic phases available. A total of 422 events were identified as aftershocks until 31 December 1969, the same period used in the hypocenter relocation. Number of events versus time shows that most earthquakes occurred during the first 10 days after the main shock (172 earthquakes).

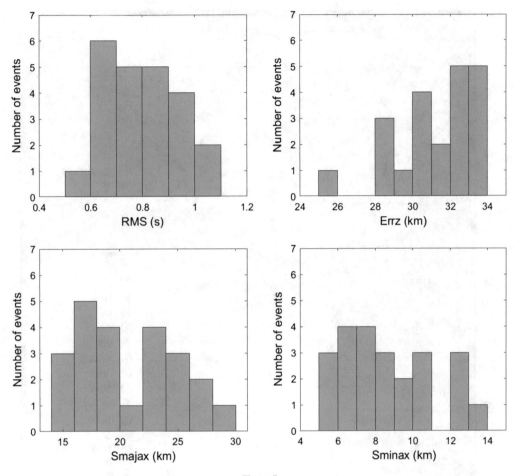

Figure 7
Histograms of errors for time origin (rms), depth (errz) and semi-axes of error elipses for epicenters

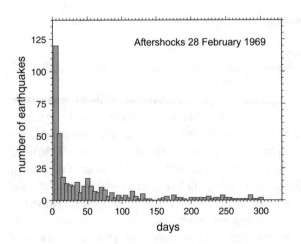

Figure 8
Number of earthquakes versus time for the 1969 seismic series

6. Comparison Between the 1969 and the 1755 Lisbon Earthquake

The 1969 earthquake occurred in the same region as the 1755 Lisbon earthquake (Fig. 1b). For the Lisbon 1755 event there are many unsolved questions such as its hypocenter location or its rupture process. However, for both earthquakes, the 1755 and 1969, we have detailed information on the intensity values in the Iberian Peninsula, so that we can compare their intensity maps and from them to draw some conclusions about the similarity or dissimilarity of their rupture process.

In Fig. 9 we have plotted the intensity values for the 1755 event in the Iberian Peninsula taken from Martínez Solares (2001) for Spain and from AHEAD,

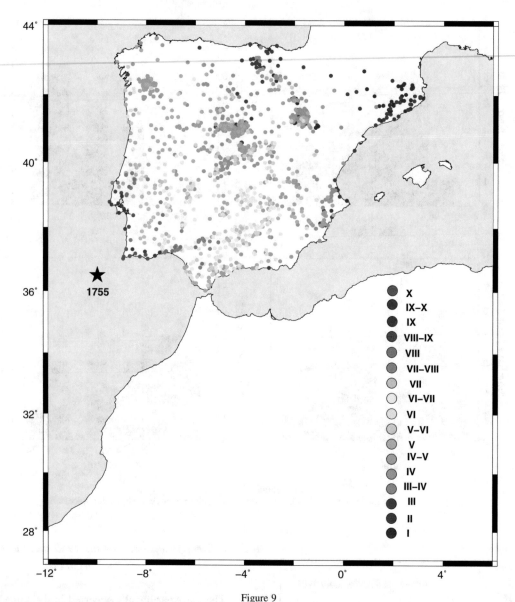

Figure 9
Intensity map for 1755 earthquake. Values from Spain are taken from Martínez-Solares (2001) and from Portugal form AHED

European Archive of Historical EArthquake Data (https://www.emidius.eu/AHEAD) for Portugal. Unfortunately no data are available for Morocco, so the comparison between both shocks will be limited to the Iberian Peninsula. When compare with the intensities distribution for the 1969 (Fig. 3) we observe important differences. For 1755 the largest intensities (X) are along the south of Portugal (the Algarve N-S coast) and in the Lisbon region, with intensities of IX and IX–X. For the 1969 event, the largest intensities are in southern Portugal, along the Algarve E-W coast (VIII to VIII–X) and lower intensities (VI–VII) in the Lisbon region). Another important difference concerns the epicenters: the 1755 event is supposed to be located (Martínez Solares and López Arroyo 2004) closer to the coast than the 1969 (150 km from Saint Vincent Cape for 1755 event versus 220 km for the 1969). However, this difference alone seems not to be sufficient to justify the different intensity pattern of the two earthquakes.

Based on these differences we explore what could have been the rupture process for the 1755 event. For the 1969 earthquake, the focal mechanism is well determined using different data and methodologies, and it corresponds to reverse faulting motion with planes oriented in E-W direction. The dimension of the rupture has been estimated 85–90 km (Fukao 1973; López-Arroyo and Udías 1972; Grimison and Chen 1988) in agreement with the spatial extension of our relocated aftershocks (Fig. 6b). López-Arroyo and Udías (1972) and Fukao (1973) have observed directivity effects using surface waves. The intensities distribution suggests that the rupture occurred along an E-W plane (Fig. 1b) propagating to the east toward the Strait of Gibraltar. This explains that the higher intensity values were reached in southern Portugal (E-W Algarve coast, Fig. 3). The different intensities distribution for the 1755 suggests also a different source rupture process. To explain this process we consider the earthquake occurred the 2009 (M_w = 5.5) very close to the proposed 1755 epicenter. For this earthquake Pro et al. (2013) have proposed that the rupture occurred in a NE-SW vertical plane propagating to the NE (in the direction from Saint Vincent Cape to Lisbon, Fig. 1b). The location and focal mechanism of the 2009 earthquake differ from those of the 1969, suggesting that the 2009 event could be an alternative, potential candidate to explain the rupture process of the 1755.

In order to check this hypothesis, we have carried out a comparison between synthetic and intensity-based PGAs for the 1755 and 1969 earthquakes. The intensity-based PGA values have been obtained using Faenza and Michelini (2010) (this relation has been used recently by the IGN (2013) to update the seismic hazard maps in Spain):

$$I = 1.68 + 2.58log(PGA) \qquad (1)$$

Using this relation, the 1755 event reached a peak value of 1.7 g, with the highest values along the Southern coast of Portugal and the western coast close to Lisbon. However, using the same relation, the largest PGAs values for 1969 earthquakes (0.4 g) were confined only in the southern coast (Algarve). In Fig. 10 we have plotted the PGA ratios r_{PGA}, defined as the ratio of the PGAs for the 1755 and 1969 earthquakes.

$$r_{PGA} = PGA_{1755}/PGA_{1969} \qquad (2)$$

We have computed r_{PGA} for each site where intensity values are available for both earthquakes within a radius of 2 km. We use warm (red) and cold (blue) colors denote respectively positive and negative anomalies of the PGA ratio with respect to its median value. Positive anomalies (red color), meaning that intensities of 1755 are larger than the corresponding of 1969, are found in the Lisbon and surrounding area, and in the N-S coast of the Algarve region, (S. Portugal) near Saint Vincent Cape (SVC). These spatial anomalies cannot be attributed to local site effects, which cancel out by considering the ratio of PGAs at the same site, and must be instead attributed to a differences in the radiation pattern of the two earthquakes, either controlled by differences in the epicentral location, depth, focal mechanism and/or directivity. This observation suggests that the location and rupture mechanism of the 1755 earthquake differed from those of the 1969 earthquake.

We used the *pyrocko* library (Heimann et al. 2017) to estimate theoretical PGA values at those sites where seismic intensities estimations are available; a similar procedure has been used by Buforn et al. (2015) and Dahm et al. (2018) for different earthquakes. Full waveform synthetic 3-components accelerations have been computed at all these sites, simulating two earthquake scenarios, using a global velocity model (AK135). In the first scenario, we simulate the 1755 earthquake assuming location, depth, magnitude and focal mechanism of the 1969 earthquake and in the second under the assumption that its location and mechanism correspond to those of the 2009 earthquake (Table 5). For both simulations, finite rupture processes are ignored, as there is no reliable information on the rupture directivity of these earthquakes. We consider in both cases a spatial point-source with 30 s rupture duration, to simulate waveforms with compatible frequency content. PGAs are here computed as the geometrical mean of the maximum acceleration of horizontal components. Figure 11 shows the ratio of the normalized PGAs computed for the 1969 and 2009 scenarios (warm and cold colors represent positive and negative deviations for the median ratio). We observe that the largest anomalies, that is, the differences between the two

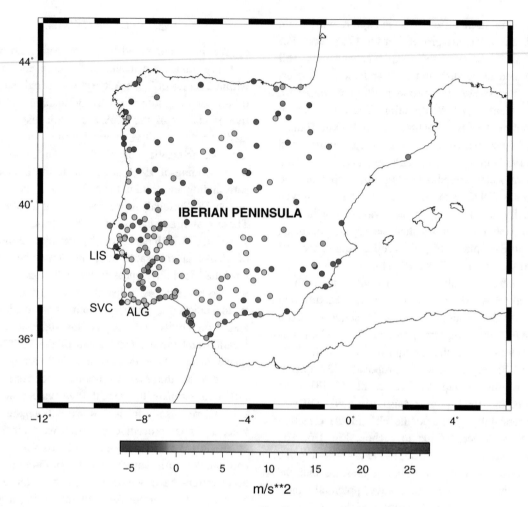

Figure 10

Map of the PGA ratio, r_{PGA}, between the 1755 and 1969 earthquake. Red color sites along the Western Portugal coast indicate anomalous large acceleration for the 1755 earthquake, with respect to the 1969 earthquake

Table 5

Focal parameters

Date	Time	Lat. (°)	Lon. (°)	h (km)	M_0 (Nm)	Str/dip/rake (°)	M_w
28-02-1969	02:40: 34	36.0469	− 10.6273	31	6×10^{20}	231/47/54	7.8
17-12-2009	01:37:49	36.4702	− 10.0318	36	2.5×10^{17}	217/89/− 58	5.5
01-01-1755	10:16:00	36.50	− 10.0	–	–	–	–

scenarios, are at the Lisbon region, with positive anomalies also present along the Western coast of Portugal. In both cases intensities deviations from the median indicate that intensities are higher for the 2009 than for the 1969 scenarios. We can then conclude that the intensity distribution of 1755 earthquake is more similar to that of the earthquake of 2009 than to the one of 1969. Therefore the distribution of intensities of the 1755 Lisbon earthquake can be better explained assuming that it took place close to the location of the 2009 earthquake and with a similar mechanism, with a reverse faulting rupture

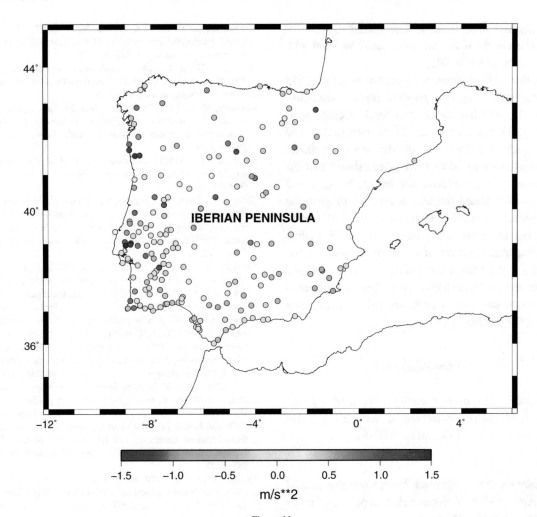

Figure 11
Map of the normalized synthetic PGA ratio for 1755 earthquake, r_{PGA}, between the 2009 and 1969 scenarios. Values are plotted as deviations from the median r_{PGA}, with warm and cold colors representing positive and negative deviations

on a nearly vertical plane in the NE-SW direction and propagating toward the NE.

7. Conclusions

- A new intensity map has been obtained for the 1969 earthquake for the Iberian Peninsula and northern Morocco using original documents some of them not used previously. The comparison between this map and the 1755 intensity map shows a different distribution; while for the 1969 event the largest intensities are found in southern Portugal (E-W coast of Algarve), for the 1755 the

maximum intensities are more to the north in the N-S coast of Algarve and in the Lisbon region (NE from the epicenter). The 1969 maximum intensities in southern Portugal can be explained by the rupture process, with reverse motion as given by its mechanism and the rupture propagating in E-W direction parallel to the coast toward the Strait of Gibraltar. This rupture is not compatible with the distribution of intensities for the 1755 earthquake.

- The relocation of the main shock and 24 larger aftershocks, show focal depths mostly between 30 and 50 km in agreement with results obtained in other studies which confirms that the seismic activity in the Saint Vincent Cape occurs in an anomalous upper mantle. The spatial distribution of aftershocks is located along an E-W band with less than 30 km width.

- Synthetic PGAs values were generated for the 1755 earthquake using two possible rupture scenarios, namely, with hypocenter and focal mechanism as the 1969 event or as the 2009 earthquake. The comparison between the distribution of synthetic and intensity-based of PGA values shows that the second scenario reproduces better the observed intensities. Based on these results we propose the location and focal mechanism of the 2009 earthquakes as better estimates for the 1755 Lisbon earthquake. However, it is important to remember that for the Lisbon earthquake the only available data are the intensities derived from documented damages and consequently all derived parameters are only estimates.

Acknowledgements

This work has been partially supported by the Spanish Ministerio de Economia, Industria y Competitividad, project CGL2017-86097-R.

Publisher's Note   Springer Nature remains neutral with regard to jurisdictional claims in published maps and institutional affiliations.

REFERENCES

Baptista, M. A., Miranda, P. M. A., Miranda, J. M., & Mendes-Victor, L. (1998). Constrains on the source of the 1755 Lisbon Tsunami inferred from numerical modelling on historical data on the source of the 1755 Lisbon Tsunami. *Journal of Geodynamics, 25,* 159–174.

Batlló, J., Carrilho, F., Alves, P., Cruz, J. & Locati, M. (2012). A new online intensity data point database for Portugal. In *Proc. 15th World Conf on Earthq Eng,* Lisbon, Portugal. Accessible online https://www.iitk.ac.in/nicee/wcee/article/WCEE2012_3683.pdf.

Buforn, E., Pro, C., Sanz de Galdeano, C., Cantavella, J. V., Cesca, S., Caldeira, B., Udías, A., & Mattesini, M. (2017). The 2016 south Alboran earthquake (M_w = 6.4): A reactivation of the Ibero-Maghrebian region? *Tectonophysics, 712–713,* 704–715. https://doi.org/10.1016/j.tecto.2017.06.033.

Buforn, E., Sanz de Galdeano, C., & Udías, A. (1995). Seismotectonics of the Ibero-Maghrebian región. *Tectonophysics, 248,* 247–265.

Buforn, E., Udías, A., & Colombás, M. A. (1988). Seismicity, source mechanism and tectonics of the Azores-Gibraltar plate boundary. *Tectonophysics, 152,* 89–118.

Buforn, E., Udías, A., Sanz de Galdeano, C., & Cesca, S. (2015). The 1748 Montesa (southeast Spain) earthquake—A singular event. *Tectonophysics, 664,* 139–153.

Carranza, M. (2016). *Sistema de Alerta Sísmica temprana para el sur de la Península Ibérica: Determinación de los parámetros de la alerta.* Doctoral Thesis. Universidad Complutense de Madrid.

Cherkaoui, T. E. (1991). *Contribution a l'etude de l'alea sismique au Maroc.* Ph.D. Dissertation Université Joseph Fourier, Grenoble.

Civiero, C., Strak, V., Custódio, S., Silveira, G., Rowlinson, N., Arroucau, P., et al. (2018). A common deep source for upper-mantle upwelling below the Ibero-western Maghreb region from teleseismic P-wave travel-time tomography. *Earth and Planetary Science Letters, 499,* 157–172.

Custodio, S., Cesca, S., & Heimman, S. (2012). Fast kinematic waveform inversion and robustness analysis: Application to the 2007 Mw 5.9 Horseshoe Abyssal Plain Earthquake Offshore Southwest Iberia. *Bulletin of the Seismological Society of America, 102,* 361–376. https://doi.org/10.1785/0120110125.

Custodio, S., Dias, N. A., Carrilho, F., Gongora, E., Rio, I., Marreiros, C., et al. (2015). Seismology Earthquakes in western Iberia: improving the understanding of lithospheric deformation in a slowly deforming region. *Geophysical Journal International, 2015*(203), 127–145. https://doi.org/10.1093/gji/ggv285GJI.

Dahm, T., Heimann, S., Funke, S., Wendt, S., Rappsilber, I., Bindi, D., et al. (2018). Seismicity in the block mountains between Halle and Leipzig, Central Germany: centroid moment tensors, ground motion simulation, and felt intensities of two M ≈ 3 earthquakes in 2015 and 2017. *Journal of Seismology, 22*(4), 985–1003.

Engdahl, R., van der Hilst, R., & Buland, R. (1998). Global teleseismic earthquake relocation with improved travel times and procedures for depth determination. *Bulletin of the Seismological Society of America, 88,* 722–743.

Faenza, L., & Michelini, M. (2010). Regression analysis of MCS intensity and ground motion parameters in Italy and its application in ShakeMap. *Geophysical Journal International, 180,* 1138–1152. https://doi.org/10.1111/j.1365-246X.2009.04467.x.

Fukao, Y. (1973). Thrust faulting at a lithosphere plate boundary: The Portugal earthquake of 1969. *Earth and Planetary Science Letters, 18,* 205–216.

Geissler, W. H., Matias, L., Stich, D., Carillho, F., Jokat, W., Monna, S., et al. (2010). Focal mechanisms for sub-crustal earthquakes in the Gulf of Cadiz from dense OBS deployment. *Geophysical Research Letters, 37,* L18309. https://doi.org/10.1029/2010GL044289.

Grandin, R., Borges, J. F., Bezzeghoud, M., Caldeira, B., & Carrilho, F. (2007). Simulations of strong ground motion in SW Iberia for the 1969 February 28 (*Ms* = 8.0) and the 1755 November 1 (*M* ∼8.5) earthquakes—II. Strong ground motion simulations. *Geophysical Journal International, 171,* 807–822.

Grevemeyer, I., Gràcia, E., Villaseñor, A., Leuchters, W., & Watts, A. B. (2015). Seismicity and active tectonics in the Alboran Sea, Western Mediterranean: Constraints from an offshore-onshore seismological network and swath bathymetry data. *Journal of Geophysical Research: Solid Earth, 120,* 8348–8365. https://doi.org/10.1002/2015JB012073.

Grimison, N. L., & Chen, W. P. (1986). The Azores Gibraltar Plate boundary: Focal mechanisms, depth of earthquakes and their tectonic implications. *Journal of Geophyical Research, 91,* 2029–2047.

Grimison, N. L., & Chen, W. P. (1988). Source mechanisms of four recent earthquakes along the Azores—Gibraltar plate boundary. *Geophysical Journal, 92,* 391–401.

Heimann, S., Kriegerowski, M., Isken, M., Cesca, S., Daout, S., Grigoli, F., Juretzek, C., Megies, T., Nooshiri, N., Steinberg, A., Sudhaus, H., Vasyura-Bathke, H., Willey, T. & Dahm, T. (2017). *Pyrocko—An open-source seismology toolbox and library.* V. 0.3. GFZ Data Services. https://doi.org/10.5880/gfz.2.1.2017.001.

IGN. (2013). *Actualización de Mapas de Peligrosidad Sísmica de España.* Madrid: Centro Nacional de Información Geográfica.

Johnston, A. (1996). Seismic moment assessment of earthquakes in stable continental regions-111. New Madrid 181 1-1812, Charleston 1886 and Lisbon 1755. *Geophysical Journal International, 1996*(126), 314–344.

Kennett, B. L. N., & Engdahl, E. R. (1991). Traveltimes for global earthquake location and phase identification. *Geophysical Journal International, 105,* 429–465. https://doi.org/10.1111/j.1365-246X.1991.tb06724.x.

Laske, G., Masters, G., Ma, Z. & Pasyanos, M., (2013). Update on CRUST1.0 A 1-degree Global Model of Earth's Crust. In *Geophys. Res. Abstracts* (vol. **15**, pp. EGU2013-2658). EGU General Assembly 2013, http://igppweb.ucsd.edu/~gabi/crust1.html.

Levret, A. (1991). The effects of the November 1, 1755 Lisbon earthquake in Morocco. *Tectonophysics, 193*(1–3), 83–94.

Lomax, A., Michelini, A. & Curtis, A. (2009) Earthquake Location, Direct, Global-Search Methods. In: R. A. Meyers (Ed.) *Encyclopedia of Complexity and System Science*, Part 5 (pp. 2449–2473). New York: Springer. https://doi.org/10.1007/978-0-387-30440-3.

Lomax, A., Virieux, J., Volant, P., & Berge-Thierry, C. (2000). Probabilistic earthquake location in 3D and layered models. In C. H. Thurber & N. Rabinowitz (Eds.), *Advances in Seismic Event Location. Modern Approaches in Geophysics* (Vol. 18). Dordrecht: Springer.

López-Arroyo, A., & Udías, A. (1972). Aftershock sequence and focal parameters of the February 28th, 1969 earthquake of the Azores-Gibraltar fracture zone. *Bulletin of the Seismological Society of America, 62*(3), 699–720.

Lozano, L., Cantavella, J. J. & Barco, J. (2019). A new 3-D P-wave velocity model for the Gulf of Cadiz and adjacent areas derived from controlled-source seismic data: Application to non-linear probabilistic relocation of moderate earthquakes. *Geophysical Journal International* (**under revision**).

Machado, F. (1966). Contribuiçao para o estudo do terramoto de 1 de Novembro de 1755, *Rev. Fac. Ciencias de Lisboa.* 2a Serie-C, XIV, fasc.1,. 19–31.

Martínez Solares, J. M. (2001). *Los efectos en España del terremoto de Lisboa.* Madrid: Instituto Geográfico Nacional.

Martínez Solares, J. M., & López Arroyo, A. (2004). The great historical 1755 earthquake. Effects and damage in Spain. *Journal of Seismology, 8,* 275–294.

Martínez Solares, J. M., Lopez Arroyo, A., & Mezcua, J. (1979). Isoseismal map of the 1755 Lisbon earthquake obtained from Spanish data. *Tectonophysics, 53,* 301–313.

Martínez-Loriente, S., Grácia, E., Bartolome, R., Sallarés, V., Connors, C., Perea, H., et al. (2013). Active deformation in old oceanic lithosphere and significance for earthquake hazard:

Seismic imaging of the Coral Patch Ridge area and neighboring abyssal plains (SW Iberian Margin). *Geochemistry, Geophysics, Geosystems, 14,* 2206–2231. https://doi.org/10.1002/ggge.

McKenzie, D. (1972). Active tectonics of the Mediterranean region. *Geophysical Journal of the Royal Astronomical Society, 30,* 109–185.

Mézcua, J. (1982). *Catálogo general de Isosistas de la Península Ibérica.* Madrid: Instituto Geográfico Nacional.

Mézcua, J, Martínez Solares, J. M. (1983). *Sismicidad del area Ibero-Mogrebí.* Publicación 203. Madrid: Instituto Geográfico Nacional.

Moreira, V. S. (1984). *Sismicidade histórica de Portugal Continental.* Lisboa: Revista Instituto Nacional de Meteorologia e Geofísica.

Moreira, V. S. (1985). Seismotectonics of Portugal and its adjacent area in the Atlantic. *Tectonophysics, 117,* 85–96.

Paula, A. O., & Oliveira, C. S. (1996). Evaluation of 1947–1993 Macroseismic information in Portugal using the EMS-92 scale. *Annali di Geofisica XXXIV, 5,* 1989.

Pro, C., Buforn, E., Bezzeghoud, M., & Udías, A. (2013). The earthquakes of 29 July 2003, 12 February 2007, and 17 December 2009 in the region of Cape Saint Vincent (SW Iberia) and their relation with the 1755 Lisbon earthquake. *Tectonophysics, 583,* 16–27. https://doi.org/10.1016/j.tecto.2012.10.010.

Sallarès, V., Martínez-Loriente, S., Prada, M., Gràcia, E., Ranero, C. R., Gutscher, M. A., et al. (2013). Seismic evidence of exhumed mantle rock basement at the Gorringe Bank and the adjacent Horseshoeand Tagus abyssal plains (SW Iberia). *Earth and Planetary Science Letters, 365,* 120–131. https://doi.org/10.1016/j.epsl.2013.01.021.

Sallarès, V., Gaillerb, A., Gutscherb, M. A., Graindorgeb, D., Bartoloméa, R., Gràcia, E., et al. (2011). Seismic evidence for the presence of Jurassic oceanic crust in the central Gulf of Cadiz (SW Iberian margin). *Earth and Planetary Science Letters, 311*(1–2), 112–123. https://doi.org/10.1016/j.epsl.2011.09.003.

Silva, S., Terrinha, P., Matias, L., Duarte, J., Roque, C., Ranero, C., et al. (2017). Micro-seismicity in the Gulf of Cadiz: Is there a link between micro-seismicity, high magnitude earthquakes and active faults? *Tectonophysics, 717*(2017), 226–241.

Stich, D., Batlló, J., Morales, J., Macià, R., & Dineva, S. (2003). Source parameters of the Mw = 6.1 1910 Adra earthquake (southern Spain). *Geophysical Journal International, 155*(2), 539–546. https://doi.org/10.1046/j.1365-246X.2003.02059.x.

Stich, D., Mancilla, F., & Morales, J. (2005). Crust-Mantle Coupling in the Gulf of Cadiz (SW-Iberia). *Geophysical Research Letters.* https://doi.org/10.1029/2005GL023098.

Stich, D., Mancilla, F., Pondrelli, S., & Morales, J. (2007). Source analysis of the February 12th 2007, Mw 6.0 Horseshoe earthquake: Implications for the 1755 Lisbon earthquake. *Geophysical Research Letters.* https://doi.org/10.1029/2007gl0300127.

Tarantola, A., & Valette, B. (1982). Generalized nonlinear inverse problems solved using the least squares criterion. *Reviews of Geophysics, 20*(2), 219.

Udías, A., & López Arroyo, A. (1970). Body and surface wave study of the source parameters of the 15, 1964 Spanish earthquake. *Tectonophysics, 9,* 323–346.

Udías, A., López Arroyo, A., & Mezcua, J. (1976). Seismotectonic of the Azores-Alboran region. *Tectonophysics, 31,* 259–289.

Vilanova, S. P., Nunes, C. F., & Fonseca, J. F. (2003). Lisbon 1755: A case of triggered onshore rupture? *Bulletin of the Seismological Society of America, 93,* 2056–2068.

Villaseñor, A., & Engdahl, R. (2005). A digital hypocenter catalog for the international seismological summary. *Seismological Research Letters, 76,* 554–559.

http://alomax.free.fr/nlloc/.

https://www.emidius.eu/AHEAD.

http://www.ign.es/web/ign/portal/sis-catalogo-terremotos.

Zitellini, N., Chierici, F., Sartori, R., & Torelli, L. (1999). The tectonic source of the 1755 Lisbon earthquake and tsunami. *Annali di Geofisica, 42,* 49–55.

(Received July 18, 2019, revised September 20, 2019, accepted September 24, 2019, Published online October 15, 2019)

Pure Appl. Geophys. 177 (2020), 1801–1808
© 2019 Springer Nature Switzerland AG
https://doi.org/10.1007/s00024-019-02368-0

Effects of the 28 February 1969 Cape Saint Vincent Earthquake on Ships

José Antonio Aparicio Florido[1] (iD)

Abstract—In the interval between the El Asnam earthquakes of 1954 and 1980, both tsunamigenic and recorded on tide gauges of the western Mediterranean coasts, the earthquake and tsunami of 28 February 1969 took place. The epicentre was located in the Atlantic Ocean, southwest of Cape St. Vincent, and tsunami waves were recorded on tide gauges of the Atlantic coasts of Portugal, Spain and Morocco. This means there were three tsunami events over a 26-year period in the current NEAMTWS coverage area. There is no evidence of land sightings of tidal wave arrivals, except for two cases in Gijón (Spain) and near the mouth of the Bou Regreg River (Morocco), which will be discussed later. The behaviour of the sea in the coastal areas was apparently normal considering the presence of a storm that was moving from the southwest of the Iberian Peninsula, accompanied by very strong rainfall in Morocco, moderate to strong rainfall in Andalusia and southern Portugal, and weak rainfall or very cloudy skies in the rest of the area. However, on the high seas, both the earthquake and the tsunami effects were clearly felt by six ships of various tonnages at different distances and azimuths from the epicentre, even with the risk of shipwreck. The crews of these ships, which at the time of the earthquake were sailing across the Atlantic near the epicentral area, reported violent vibrations, very high waves and specific damages, which call into question the belief that seaquakes and tsunamis are not perceptible on deep waters, inviting us to analyse the risk posed by earthquakes with submarine epicentres for maritime navigation.

Keywords: 28 February 1969 earthquake, effects on ships, ida Knutsen, esso Newcastle, manuel Alfredo, tide gauges, tsunamis, hogging and sagging efforts, gorringe Bank.

1. Seismic and Tsunamigenic Context

In the interval between the two Algerian earthquakes of El Asnam in 1954 and 1980, both tsunamigenic, a widely felt earthquake took place in the Iberian Peninsula and Morocco, reminiscent of the devastating Lisbon earthquake of 1 November 1755. It happened at

02.40.32 UTC (t_0) on Saturday, 28 February 1969, at coordinates 35.98° N 10.81° W. Its location in the Horseshoe Abyssal Plain, just 200 km southwest of Cape Saint Vincent, and the high magnitude of 7.8 M_w (USGS, IGN) caused seismic waves to hit Portugal, Spain and Morocco, where widespread damage was recorded, with catastrophic results in some cases, causing scenes of panic among the population. But the most interesting thing is that it generated a "small" tsunami that, although not perceptible everywhere and to almost no one near the coast, left data records in tide gauges of the three countries reached by the waves: 93.2 cm in Cascais, 84.3 cm in Lagos, 28.4 cm in Cádiz (Baptista et al. 1992), 12 cm in A Coruña, 30 cm in Chipiona, 0.17 in Santa Cruz de Tenerife and 1.2 m in Casablanca (Martínez Solares 2005).

The generation of this non-destructive tsunami in the area under the control of the current Northeastern Atlantic and Mediterranean Tsunami Warning System (NEAMTWS) is part of the historical context of other tsunami events that for a few years reached several European and African coasts. One of them occurred on 9 September 1954, of magnitude $M_D = 6.7$ and intensity X-XI (IGN), with the epicentre in the Algerian city of Chlef (El Asnam), killing 1243 people. Being located 30 km inland, it was able to generate a tidal wave registered by the tide gauge of Alicante with wave heights of 55 cm (Martínez Solares 2005) and by the tide gauges of Málaga, Algeciras and Ceuta with maximum wave amplitude of 33 cm for the first one and 5–7 cm for the last two (Soloviev et al. 2000).

A new earthquake took place on 10 October 1980 in the same epicentral area of El Asnam, with a magnitude $M_s = 7.3$ and intensity IX (USGS), which left around 5000 dead and several cities devastated. The fault plane was located 45 km south of the Mediterranean coast and 15 km close to the city of

[1] Instituto Español para la Reducción de los Desastres, c / Rondeñas, 10 - 3° B, 11100 San Fernando, Cádiz, Spain. E-mail: aparicioflorido@gmail.com

Beni Rached, where the previous one had been registered in 1954. Again, tsunami waves of up to 70 cm were registered southeast of the Iberian Peninsula by the tide gauges of Alicante, Cartagena, Almería, Málaga and Algeciras (Soloviev et al. 2000). The waves arrived in Alicante 50 min later and in Algeciras 1 h 30 min later. However, west of Gibraltar Strait or in Palma de Mallorca there are no records.

Therefore, the 1969 tsunami was the second of this short sequence in the western NEAM area between 1954 and 1980; that is, three tsunami events in 26 years could be catalogued thanks to seismic, oceanographic and tide gauge instrumentation and to the combination of the respective data records. Prior to these, there were two other earthquakes, but with epicentres far further from the Iberian Peninsula and Morocco: the first one in 1929 recorded in Lagos and the Azores Islands (Fine et al. 2005, with data provided by Miranda) and the second in 1941 recorded in Casablanca and Essaouira (Morocco) and Cascais, Lagos, Madeira and Ponta Delgada (Portugal) (Kaabouben et al. 2009). The 1941 tsunami, generated by a submarine landslide, was located in Grand Banks outside the NEAMTWS coverage area. If they had happened in the pre-instrumental period, perhaps only that of 1969 would have been included in the historical catalogues because of the unmistakable effects it had on maritime navigation. A strange case happened in 1983, when the Spanish authorities issued a public tsunami warning on the coast of Cádiz (no earthquakes were recorded in nearby areas) that forced the evacuation of some beaches (Aparicio Florido 2017). This reasoning leads us to think about the long list of *invisible* tsunamis that may have taken place in the past but were not recorded in the literature.

2. Effects on Ships at Sea

We have evidence of at least six ships positioned in the northeastern Atlantic waters in the current NEAMTWS coverage area, which felt, suffered or observed the effects of P-waves and tsunami waves released by the earthquake: five in latitudes north of the epicentre and one to the south (Fig. 1).

The first was the "Ida Knutsen" (not "Knudsen"; see Hammerborg 2011), a 200-m-long Norwegian tanker with 32,580 tonnes deadweight, built in 1958 and owned by the shipbuilder D/S A/S Jeanette Skinner; it was sailing in ballast (without cargo) from Portugal to the Persian Gulf at the coordinates 36.12° N 10.70° W at a distance of 20 km and azimuth 029° from the seismic focus. The vertical impact received under its hull for just 10 s was very strong (Ambraseys 1985), accompanied by a lifting force that seemed to raise it from the sea and drop it suddenly back into the water in a single shaking that caused severe structural damage. This first upward movement was not related to the weather conditions, since the sea was calm and the prevailing wind was weak (Jia 2017), so the violent vertical movement was absolutely disproportionate. The dangerous conditions, having the shaft of one of the propellers misaligned, forced the crew to struggle back to the port of departure and check the damage in a Portuguese shipyard. The ship showed a longitudinal bending of the hull, mainly in the air draft, with the hull plates, beams and internal bulkheads cracked or deformed and filtrations in the tanks; the radio, radar and navigation instruments had been rendered useless, and the china, doors and furniture had been smashed. The fact that it was sailing in ballast may have increased the severity of the damage by offering less resistance to the torsion and compression of the materials and being more vulnerable to the efforts of "hogging" and "sagging" (Fig. 2), favoured by a lower weight in the castle of the bow, which was the part of the ship that rose the most. Hogging and sagging stresses occur when the length of a sea wave is equal or similar to the length of a ship. If the ship is located over the crest of the wave (hogging), the centre of the hull receives a greater lifting motion (Fig. 2b); if it is located over the trough of the wave (sagging), the lifting motions are concentrated in the bow and stern (Fig. 2c). In both cases, the bending of the ship could cause the hull to snap or crack, leading to its wreckage. It should be noted that, under the influence of abnormal wave heights, the largest sagging moments usually occur after the large wave crest has passed through the midship and the bow has dived into the following wave crest (Fonseca et al. 2006), as was the case for the "Ida Knutsen", located just above the place where the tsunami wave was generated. Although Ambraseys (1985) considers that

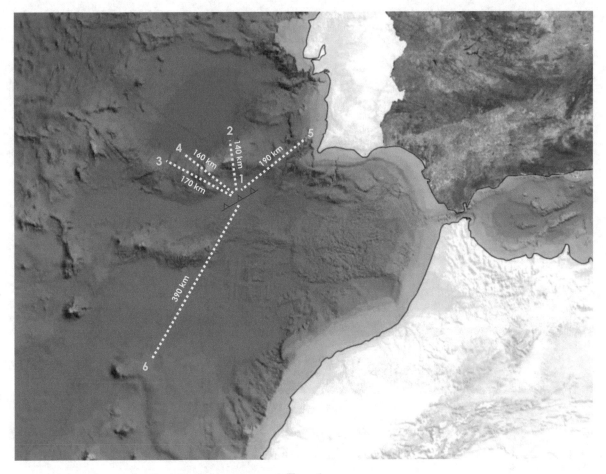

Figure 1
Position of the six vessels that felt the effects of the earthquake and tsunami of 28 February 1969: (1) Ida Knutsen; (2) Manuel Alfredo; (3) Esso Newcastle; (4) Arapiles; (5) Toubkal; (6) Diana. Source: self-made

all of the damage to this ship was caused by the impact of P-waves, it cannot be ruled out that part of it, especially the bending effects on the hull (Malisan et al. 2012), were due to a combination of P-waves and the sagging and hogging effects until stabilization of the sea surface over the epicentral area. The dry dock inspection of the "Ida Knutsen" determined the total loss of the ship, although 2 months later it was sold "as is" to the Greek shipowner Achilles Halcoussis for $730,000; he repaired it and renamed the ship "Petros Hajikyriakos".

The second ship closest to the epicentre was the "Manuel Alfredo", a 103-m-long cargo and passenger ship of 3600 tonnes deadweight, built in 1954, flying the Portuguese flag with capacity for 92 passengers and 62 crew members, sailing between

Angola and Lisbon. "Manuel Alfredo" was located on the coordinates 37.26° N 11.00° W at 140 km and azimuth 232° from the epicentre, when it was strongly shaken, followed by other vibrations of similar intensity, without this strange phenomenon being related to an earthquake. Initially, it was thought that one of the machines had stopped or that the blade of one of the propellers had been broken. The sensation was like touching a rocky bottom or climbing stairs, but at depths of 4000 m that was impossible. Even so, the third pilot, João António Pereira Chuva, who was on duty at that time, ordered stopping the machines and checking the ship, but everything was in perfect condition. The episode lasted a few seconds, not even half a minute. From the deck, the sea looked abnormally rough. Pereira

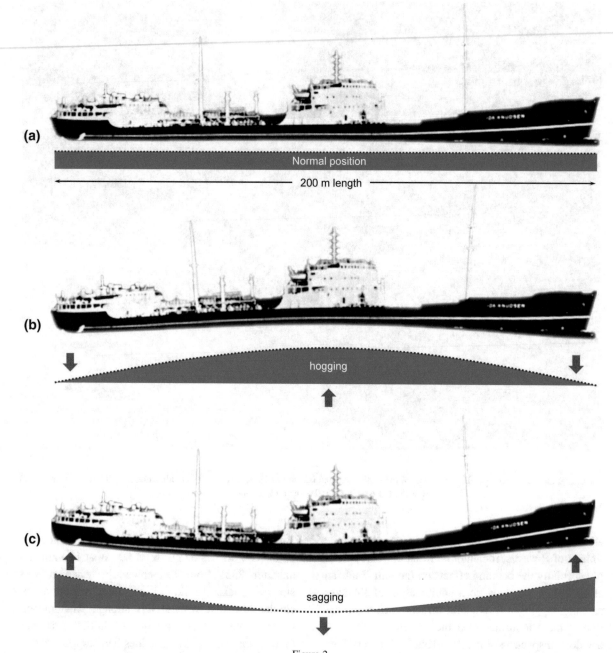

Figure 2
Modelling of hogging and sagging moments using the tanker Ida Knutsen's shape as reference: **a** normal floating position; **b** hogging effort; **c** sagging effort. Source: self-made

saw very high waves covering the ship and submerging the bow. It was a chestnut-colured water, like "mud foam", that behaved in a very unusual way. Even the crew, accustomed to withstanding the movement of storms, fell to the ground. The passengers were sleeping in their cabins and no one panicked. The few people who woke up supposed it was a breakdown and went back to bed. Half an hour later, they heard from the radio news that it had been an earthquake. None of the ship's crew had ever experienced the effects of seismic waves at sea before. "If it had lasted more than a minute, I think

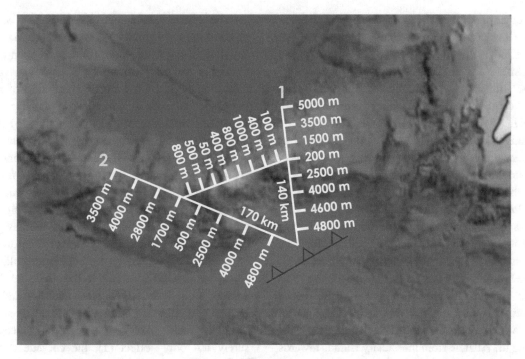

Figure 3
Bathymetric data of the Horseshoe Abyssal Plain and Gorringe Bank from the epicentre to the position of Manuel Alfredo (1) and Esso Newcastle (2). Source: self-made

the ship would have had many risks of suffering serious damage, even going to the bottom", said João Pereira. He added that for hours abnormally large, invisible and sudden waves swayed the ship, causing it to remain motionless at the end of each crest, before turning to the other side in a cyclic movement, which reinforces our theory that the ships near the epicentre and around the Gorringe Bank, like "Ida Knutsen", were subjected to the effects of hogging and sagging.

A third vessel that noticed the arrival of seismic waves was the "Esso Newcastle", a modern tanker for its time, of 51,429 tonnes deadweight and 226.5 m length, built in 1963 by Vickers-Armstrong in the Newcastle shipyards for Esso Petroleum. Given its measurements, it was not exactly a "small" ship; however, it seemed to be a light object adrift. At the time of the event it was located at position 36.52° N 12.55° W, at 170 km and azimuth 291° from the epicentre. Used to struggling in rough seas in the face of stormy waves that at other times had covered the deck of the ship from side to side, this incident could not be compared with any other critical situation

experienced previously. Despite being nine times more distant than "Ida Knutsen"—which was just above the seismic focus—the crew felt the effect of a vertical movement from bottom to top, in the same way as it was perceived by the Norwegian tanker and similar to the stair-climbing feeling of the Portuguese passenger ship; they believed that they had collided with a floating object. However, the people on all three ships described thinking that something had happened to the propellers and, as happened to "Manuel Alfredo", that they had lost a blade.

Very near the "Esso Newcastle" was the Spanish flag vessel "Arapiles", an oil tanker owned by the CEPSA Company, with 12,500 tonnes deadweight and 8368 gross register, 142.2 m in length, built in 1931 and scrapped in 1971. Shortly after the earthquake took place, the Real Observatorio de Marina of San Fernando (Spain) received a message from the captain of this ship stating that, sailing over coordinates 36.68 N 12.28 W, at 160 km from the epicentre, the crew perceived "very strong vibrations" at 03:45 local time (02:45 UTC); unfortunately, in this case we have no information about other effects related to the tsunami or the sea-quake. However, at a similar position, the effects felt were similar to those reported by the "Esso Newcastle".

The fifth ship was "Toubkal", a 150.3-m-long general cargo ship of 12,950 tonnes deadweight, built in 1961, owned by the Compagnie Marocaine de Navigation (COMANAV), used to the transport of phosphates, cereals and Cuban sugar between Morocco and the ports of Western Europe. At the time of the earthquake, it was located at coordinates 37.18° N 9.20° W, 190 km and azimuth 046° from the epicentre and about 90 km from the port of Sines. We only know that it felt "violent vibrations" for 60 s (Ambraseys 1985) without suffering any damage.

The last of the six vessels studied here is the 75.5-m-long corvette "Diana" (F-63), with 1136 tonnes deadweight, built in 1955, owned by the Spanish Navy, which at the time of the earthquake was crossing the Atlantic from the Canary Islands towards Cádiz, ruling the sea, to make navigation more comfortable, which means that they had the bow pointed slightly to the west, as if they were sailing to the Azores and not to Spain. The ship had just completed a joint exercise with the French Navy called Atlántide-69 and was returning to its base with enough time, so instead of heading 042°—which would make the trip very uncomfortable for the crew—they set course at 035°. The positioning of the ship is imprecise, but the officer on duty, Francisco José Súnico Varela, remembered that they were halfway between the Canary Islands and Cádiz, so approximately at coordinates 32.80° N 12,50° W, about 390 km and azimuth 205° to the south of the epicentre. Even so, the crew thought they had become stuck on a shoal. It was a very strange sensation, as if the boat were being shaken by a powerful force accompanied by a deep, dull, non-metallic sound coming from inside the vessel and not from outside, like when someone shakes a glass jar full of sand. This agitation lasted about 4 s. After inspecting the ship completely, they did not detect any damage or failure in the engine room, so they continued the route without incident. They did not receive any warning or news about the circumstances that could have caused the phenomenon, so they never related the episode to the 28 February 1969 earthquake.

The effects felt in these six ships can be summarized as: (1) agitation, shaking or beating of the structure of the whole ship; (2) the sensation of bottoming or colliding with a rock or sand shoal or with a floating object; (3) acoustic perception of a dull noise; (4) damage or sensation of damage to the propellers, rudder or propulsion systems; (5) swinging that sometimes awakened people who were asleep or making it difficult to stand up; (6) rising or a sensation that the ship was rising vertically from the bottom to top above the surface of the water; (7) falling objects and damage, hitting or moving furniture and interior doors; (8) damage to navigation control or communications systems; (9) deformation or openings of the hull plates or interior bulkheads. The list is ordered according to the effects felt by the vessels, from lowest to highest intensity and from furthest to closest epicentral distance, so that "Toubkal" and "Arapiles" perceived—at least, but surely not only—effect (1), the corvette "Diana" (1–3), "Manuel Alfredo" and "Esso Newcastle" (1–4) and "Ida Knutsen" (1–9). It is not possible, however, to establish a clear correlation between these effects and those described on macroseismic scales, because certain effects may have been a consequence of tsunami waves or a combined effect. Effects (1–4) and (6) were exclusively the effect of P-waves, but the rest may have been the result of a double component: P-waves and tsunami waves. However, the effects felt by the corvette "Diana" could correspond to intensity III–IV (EMS-98), "Toubkal" intensity IV, "Manuel Alfredo" and "Esso Newcastle" intensity V–VI and "Ida Knutsen" intensity VIII, although the tonnage, length, construction materials and other characteristics of the ships were surely closely related to how and how long the seismic waves were felt. Therefore, these different characteristics should be part of a new classification of ships' vulnerability, like for buildings, because the EMS-98 scale or any other intensity scale currently in use is only relevant for buildings, people, animals and the natural environment on land.

There was one last testimony offered by fishermen who used to fish off the coast of Matosinhos, Portugal, 600 km north of the epicentre. On 29 March 1969, Diário da Manhã published that these fishermen, from their respective boats, had felt a sudden and strong disturbance of the sea without understanding the reason for the abrupt change, but no more information is available for a more extended analysis.

Most of the effects described above are consequences of the propagation and inner resonance of P-waves when they cross the hull of a ship, but some effects are closely related to the tsunami waves. As described, the "Ida Knutsen" suffered a deformation of its structure due to the impact of P-waves or effects of hogging and sagging, or their combination, while in the case of the "Esso Newcastle" and "Manuel Alfredo", the waves covered the deck. Contrarily, these strong oscillations of the sea were not observed by the "Toubkal" or "Diana". The captain of the "Manuel Alfredo" described them as "enormous waves" that "enveloped" the ship. These observations do not agree with the general theory in which tsunami wave amplitudes offshore are very small, no more than a few centimetres, and not perceptible by ships in deep waters. When the earthquake happened, the "Ida Knutsen", on the one hand, and the "Esso Newcastle" and "Manuel Alfredo", on the other, were located on both sides of the Gorringe Bank, sailing at depths of 4800–5000 m, but between the epicentre (4800 m deep) and the "Esso Newcastle" (3500 m deep) and "Manuel Alfredo" (5000 m deep), the Gorringe Bank crest, parallel to the fault plane, sharply decreases those depths up to < 100 m (Fig. 3). A cross-section of this bathymetric alteration is shown in Fukao (1973). The coincidence of these effects near the Gorringe Bank, but not far from this geological structure, leads us to consider the possible cause-effect relationship for future research.

3. Coastal Observations of the Tsunami

In addition to the unusual behaviour of the sea described by the crew members of the ships mentioned above, we have reports of two coastal sightings in Rabat (Morocco) and Gijón (Spain). In the Bou Regreg River, 20 km from Rabat, next to the Grou dam, under construction in 1969, an incoming tidal current was observed. A 3D video-based mapping model made by Amine et al. (2018) estimates a maximum water elevation that did not exceed 1 m

inside the estuary and decreased below 0.5 m just 1 km upstream of the estuary. A more uncertain case happened near the coast of Gijón, where small boats broke their moorings and drifted, according to the Fiel news agency.

4. Conclusions

Fifty years after the earthquake and tsunami of 28 February 1969, information about the effects on ships has been scarce and poorly analysed, with some exceptions such as Ambraseys (1985). The names of these already known vessels are the "Ida Knutsen", "Esso Newcastle", "Manuel Alfredo" and "Toubkal". This article makes a deeper analysis of the previous ones and provides information for two new ships, both under the Spanish flag: the tanker "Arapiles" and the corvette "Diana". In particular, the direct testimony obtained from a crew member of the corvette "Diana" gives us a good description of how the P-waves were felt inside the ship when they were sailing around 400 km from the epicentre.

All the ships mentioned here perceived the seaquake shaking, but three of them also described certain effects clearly related to the tsunami waves: the "Ida Knutsen", "Esso Newcastle" and "Manuel Alfredo". These three ships are precisely those that were closest to the epicentral area and in the surroundings of the Gorringe Bank, more properly called Gorringe Ridge, where the bathymetry changes suddenly from 5000 m to a minimum depth of ∼ 25/30 m below sea level (Auzende et al. 1984; De Alteriis et al. 2003; Jiménez-Munt et al. 2010). The effects of both P-waves and tsunami waves described above can be summarized in a grading scale (1–9) similar to the macroseismic intensity scales currently in use (Musson et al. 2010).

Concerning the tsunami effects and the initial upward waves on ships over the epicentral area, we must also consider the effects of hogging and sagging, which can cause the bow and stern to rise more than the central part of the hull (sagging) or the centre to rise more than the bow and stern (hogging). The resulting effects are conditioned by the length of the ship, the quality and flexibility of the materials and the cargo. The most prominent case is the Norwegian tanker "Ida Knutsen", which was sailing in ballast and suffered a longitudinal bending of the hull that meant the total loss of the ship.

Some other ship crews have reported seismic noise and shaking while sailing over deep waters around the world. All the effects they described are similar to those felt by the crews of the six ships studied in this work. However, a larger comparative case study is necessary to better understand these observations and results.

Publisher's Note Springer Nature remains neutral with regard to jurisdictional claims in published maps and institutional affiliations.

REFERENCES

Ambraseys, N. (1985). A damaging seaquake. *Earthquake Engineering and Structural Dynamics, 13,* 421–424. https://doi.org/10.1002/eqe.4290130311.

Amine, M., Ouadif, L., Bahi, L., Baba, K., & Astier, S. (2018). Establishment of process of tsunami risk mapping modelling in a 3D video for the city of Rabat (Morocco): Application to the worst scenario 1755 and the moderate scenario 1969. *International Journal of Civil Engineering and Technology, 9*(9), 1559–1572. **(ISSN: 0976-6316)**.

Aparicio Florido, J. A. (2017). *1755: el maremoto que viene* (p. 251). Cádiz: Q-book. **(in Spanish, ISBN: 978-84-15744-50-4)**.

Auzende, J. M., Ceuleneer, G., Cornen, G., Juteau, T., Lagabrielle, Y., Lensch, G., et al. (1984). Intraoceanic tectonism on the Gorringe Bank: Observations by submersible. In G. Gass, S. J. Lippard, & A. W. Shelton (Eds.), *Ophiolites and oceanic lithosphere* (Vol. 13, pp. 113–120). London: Geological Society Special Publications. https://doi.org/10.1144/gsl.sp.1984.013.01.10.

Baptista, M. A., Miranda, P. M. A., & Victor, L. M. (1992). Maximum entropy analysis of Portuguese tsunami data: the tsunamis of 28.02.1969 and 26.05.1975. *Science of Tsunami Hazards, 10,* 9–20. **(ISSN: 0736-5306)**.

De Alteriis, G., Passaro, S., & Tonielli, R. (2003). New, high resolution swath bathymetry of Gettysburg and Ormonde Seamounts (Gorringe Bank, eastern Atlantic) and first geological results. *Marine Geophysical Researches, 24*(3–4), 223–244. https://doi.org/10.1007/s11001-004-5884-2.

Fine, I. V., Rabinovich, A. B., Bornhold, B. D., Thomson, R. E., & Kulikov, E. A. (2005). The Grand Banks landslide-generated tsunami of November 18, 1929: Preliminary analysis and numerical modelling. *Marine Geology, 215,* 45–57. https://doi.org/10.1016/j.margeo.2004.11.007.

Fonseca, N., Soares, C. G., & Pascoal, R. (2006). Structural loads induced in a containership by abnormal wave conditions. *Journal of Marine Science and Technology, 11,* 245–259. https://doi.org/10.1007/s00773-006-0222-9.

Fukao, Y. (1973). Thrust faulting at lithospheric plate boundary: The Portugal earthquake of 1969. *Earth and Planetary Science Letters, 18,* 205–216.

Hammerborg, M. (2011). Inheriting strategies: understanding different approaches to shipping during the World War I boom in Haugesund, Norway. In L. R. Fischer & E. Lange (Eds.), *New directions in Norwegian maritime history* (Vol. 13, pp. 113–198). St. John's: International Maritime Economic History Association **(ISBN: 978-0-9864973-6-0)**.

Jia, J. (2017). *Modern Earthquake engineering: Offshore and land-bases structures* (p. 848). Berlin: Springer. https://doi.org/10.1007/978-3-31854-2.**(ISBN: 978-3-642-31854-2)**.

Jiménez-Munt, I., Fernández, M., Vergés, J., Afonso, J. C., García-Castellanos, D., & Fullea, J. (2010). Lithospheric structure of the Gorringe Bank: Insights into its origin and tectonic evolution. *Tectonics, 29*(5), TC5049. https://doi.org/10.1029/2009tc002458.

Kaabouben, F., Baptista, M. A., Iben Brahim, A., El Mouraouah, A., & Toto, A. (2009). On the Moroccan tsunami catalogue. *Natural Hazards and Earth System Sciences, 9,* 1227–1236.

Malisan, J., Jinca, M. Y., Parung, H., & Saleng, A. (2012). Strength analysis of traditional ships in efforts to improve sea transportations safety in Indonesia. *International Journal of Engineering & Technology, 12*(5), 118–123.

Martínez Solares, J. M. (2005). Tsunamis en el contexto de la Península Ibérica y del Mediterráneo. *Enseñanzas de las Ciencias de la Tierra, 13*(1), 52–59. **(ISSN: 1132-9157. In Spanish)**.

Musson, R. M. W., Grünthal, G., & Stucchi, M. (2010). The comparison of macroseismic intensity scales. *Journal of Seismology, 14*(2), 413–428. https://doi.org/10.1007/s10950-009-9172-0.

Soloviev, S. L., Solovieva, O. N., Go, C. N., Kim, K. S., & Shchetnikov, N. A. (2000). *Tsunamis in the Mediterranean Sea 2000 b.C.-2000 a.D* (p. 237). Berlin: Springer. **(ISBN: 978-90-481-5557-6)**.

(Received May 24, 2019, revised November 1, 2019, accepted November 12, 2019, Published online November 27, 2019)

Pure Appl. Geophys. 177 (2020), 1809–1829
© 2020 Springer Nature Switzerland AG
https://doi.org/10.1007/s00024-019-02401-2

Study of the PGV, Strong Motion and Intensity Distribution of the February 1969 (Ms 8.0) Offshore Cape St. Vincent (Portugal) Earthquake Using Synthetic Ground Velocities

C. Pro,[1] ⓘ E. Buforn,[2,3] A. Udías,[2] J. Borges,[4] and C. S. Oliveira[5]

Abstract—The 28 February 1969 (Ms 8.0) Cape St. Vincent earthquake is the largest shock to have occurred in the region after the Lisbon earthquake of 1755. However, the study of the rupture process has been limited due to the characteristics of the available seismic data which were analogue records that were generally saturated at both regional and teleseismic distances. Indeed, these data consist of just one accelerograph record at the 25th April Bridge in Lisbon (Portugal) and the observed intensities in the Iberian Peninsula and northern part of Morocco. We have used these data to simulate the distribution of PGV (Peak Ground Velocity) for the 1969 event at regional distances (less than 600 km) by using a 3D velocity model. The PGV values are very important in seismic hazard studies. The velocity model and the methodological approach were tested by comparing synthetic and observed ground velocities at regional distances for two recent, well-studied earthquakes that occurred in this region, namely, the 2007 (Mw = 5.9) and the 2009 (Mw = 5.5) earthquakes. By comparing the synthetic and observed PGA (Peak Ground Acceleration) at Lisbon, the focal depth was estimated equal to 25 km and the seismic moment equal to 6.4×10^{20} N m (Mw = 7.8) for 1969 earthquake. With these parameters, PGV values were obtained for 159 sites located in the Iberian Peninsula and northern region of Morocco where we have felt intensity values. Using different empirical relations, the instrumental intensity values were calculated and compared with the felt intensities. As a result, the synthetic PGV values obtained in this study for the 1969 earthquake could be used as reference values, and the methodological approach would allow the PGV and intensity to be simulated for other events in the region.

Keywords: 1969 Cape St. Vincent earthquake, PGV values, instrumental intensity, synthetic ground motion.

1. Introduction

The 28 February 1969 (Ms 8.0) earthquake shocked the Iberian Peninsula, Morocco and south to the Canary Islands, and was followed by a long sequence of aftershocks continuing for at least 10 months (Lopez Arroyo and Udías 1972). The epicentral region, located in the Atlantic Ocean offshore of Cape St. Vincent, the same as that of the famous Lisbon earthquake of 1755, is characterized by a very long return periods for high-magnitude (Mw ≥ 8) earthquakes and tsunamis (Cunha et al. 2012) and the epicentres of several large historical earthquakes have been hypothetically located in this zone (Udías et al. 1976) (Fig. 1). Besides the 1755 Lisbon earthquake (I_{max} = X), there is information of large historical earthquakes in 880, 1356, 1531 and 1761 with similar characteristics (Buforn et al. 1988b; Baptista et al. 2006). The rupture process for the 1755 Lisbon earthquake is still a matter of debate, with several models having been proposed (Baptista et al. 2003; Grandin et al. 2007b; Gutscher et al. 2006; Stich et al. 2007; Pro et al. 2013; Vilanova et al. 2003; Zitellini et al. 2001). But all authors agree that there is a clear possibility of a future occurrence of a similar very large earthquake in this zone.

The study of the 1969 earthquake is limited by the nature of the data: analogue and completely saturated seismograms at regional distances. Body waves at teleseismic distances also are saturated. The only non-saturated record at regional distance is the ground acceleration record from a SMAC-type accelerometer (Strong Motion Acceleration Committee, a model first developed in Japan in 1953), located at the 25th April Bridge in Lisbon (Portugal). Another available data at regional distances are the

[1] Dpto. de Física- Centro Universitario de Mérida, Universidad de Extremadura, 06800 Mérida, Badajoz, Spain. E-mail: cpro@unex.es
[2] Dpto. de Física de la Tierra y Astrofísica, Universidad Complutense, 28040 Madrid, Spain.
[3] IGEO, Universidad Complutense, CSIC, Madrid, Spain.
[4] Dpto. de Física, Escola de Ciências e Tecnologia, Instituto Ciências da Terra, Universidade de Évora, Évora, Portugal.
[5] Instituto Superior Técnico, CERis, Universidade de Lisboa, Lisboa, Portugal.

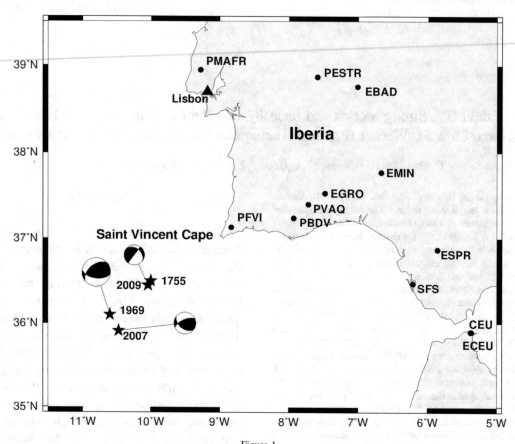

Figure 1
Broadband seismic stations used to test the velocity model. Stars: epicentre location for the 1755, 1969, 2007 and 2009 earthquakes. Triangle: acceleration data at 25th April Bridge (Lisbon)

observed intensities in the Iberian Peninsula and northern part of Morocco (López-Sánchez et al. 2019). With these data, following the procedure shown in Fig. 2, we obtained the distribution of the PGV (Peak Ground Velocity) values, which are very important for seismic hazard studies. In STEP 1, we checked the methodological approach and the 3D crustal and upper mantle model used by comparing synthetic and recorded ground velocities at regional distances (less than 600 km) for the two recent and well-studied earthquakes, of lower magnitude, that have occurred in this region, those of 2007 (Mw = 5.9) and 2009 (Mw = 5.5). For these earthquakes one has broad-band records at regional distances, so that we could compare the observed ground velocity data with the synthetic values generated with the 3D velocity model, and thus validate its use for this region. In STEP 2 we generated synthetic PGA (peak

ground acceleration) values for the 1969 earthquake using the tested 3D velocity model, and compared them with the PGA value observed at Lisbon obtained from the accelerograph record. This step is important because it allows some of the source parameters to be checked and fixed upon. With these fixed parameters we proceeded to obtain PGV values from the synthetic velocity records generated at regional distances (less than 600 km) for sites where observed intensities were available for the 1969 earthquake. Finally, we estimated instrumental intensities (I_{MM}) from the synthetic PGV values and compared them with the observed ones. To obtain I_{MM} from PGV, we used the empirical relations proposed by Wald et al. (1999), Grandin et al. (2007b), and Faenza and Michelini (2010), and a specific relation for the Cape St Vincent region that we established in this present study.

DATA:

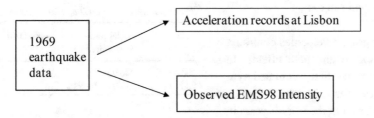

STEP 1: 3D velocity model validation

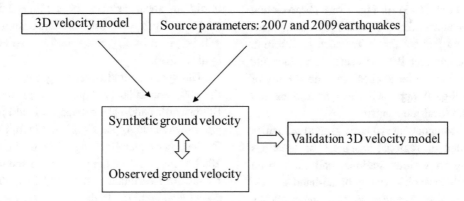

STEP 2: synthetic PGV and empirical relations validation for 1969 earthquake

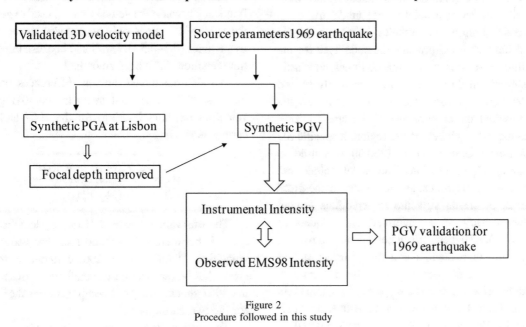

Figure 2
Procedure followed in this study

2. Synthetic Ground Velocity for the Offshore Cape St. Vincent Region

To generate synthetic ground velocities correctly, we should include the source and path effects. In order to solve this problem, we first calculated synthetic ground velocities for the two largest earthquakes that have occurred in the last 40 years in the offshore Cape St. Vincent region, namely, those of 12 February 2007 (Mw = 5.9) and 17 December 2009 (Mw = 5.5), with very well-known rupture process (Pro et al. 2013). In Fig. 1 one observes that the focal mechanism for 2007 and 1969 earthquakes are similar, but that for the 2009 shock is different. However, we consider it to be useful to include the 2009 earthquake in order to check that the validity of the methodological approach developed does not depend on the focal mechanism.

The three-components synthetic ground velocities were computed at regional distances at locations corresponding to various stations, and have been compared with those observed by broad-band seismic stations of the IGN, Western Mediterranean (WM), and Instituto Português do Mar e da Atmosfera (IPMA) networks (Fig. 1). The synthetic values were obtained considering a point-source model using the E3D code (Larsen and Schultz 1995), an elastic finite-difference wave propagation code used for the modeling of seismic waves that assumes an elastic and isotropic medium. The grid is regularly spaced and the variables computed at each node are the velocities and the components of the stress tensor. Due to the heterogeneity of the region, a specific 3D crustal model Grandin et al. (2007a) was used to generate the Green functions. The model includes the major seismogenic zones, and accounts for the gross properties of crustal structure to reproduce ground motion at low and intermediate frequencies (f < 0.5 Hz). Higher frequencies are, however, linked with a more detailed structure of the crust, and are poorly constrained over most of the area studied. To characterize the source, we used the parameters listed in Table 1 and the source time functions calculated by Pro et al. (2013), low pass filtered at 0.5 Hz. We took a grid spacing equal to 0.5 km in all the directions. There is a limitation in the frequencies of synthetic seismograms because, in order to

Table 1

Source parameters for 2007 and 2009 earthquake (Pro et al. 2013)

Date	Nodal planes Strike/dip/rake	Depth (km)	M_0 (N°m)	M_w
2007/2/12	246°/64°/51°	30	8.1×10^{17}	5.9
2009/12/17	217°/89°/− 58°	36	2.5×10^{17}	5.5

minimize numerical dispersion, the shortest wavelength must be sampled at five grid-points per wavelength (Levander 1988). In this study, with a minimum shear velocity of c_min = 1.5 km/s and grid spacing h = 0.5 km, the maximum frequency will be given by f_{max} = c_min/5 h= 0.6 Hz (Grandin et al. 2007a).

The synthetic and observed ground velocities for the 2007 and 2009 earthquakes were filtered with a Butterworth band-pass between 0.1 and 0.4 Hz, and are shown in the Appendix (Figs. 12a, b, 13a, b, 14a, b for 2007 earthquake and Figs. 15a, b,16a, b,17a, b for 2009 event). As an example, Fig. 3a shows the result for the 2007 earthquake at the PFVI station in the NS component and Fig. 3b the vertical component for the 2009 earthquake at the PMAFR seismic station. The two show reasonable agreement in amplitude but not in frequency because the velocity model used to generate the synthetics values only reproduces those of low frequencies (0.1–0.5 Hz) well and because the source time functions have been smoothed.

We also calculated the rms differences between the observed (ob(i)) and synthetic (syn(i)) ground velocities (without filtering) for the 2007 and 2009 earthquakes using the equation:

$$rms = \frac{\sqrt{\sum (ob(i) - syn(i))^2}}{\sqrt{\sum (ob(i))^2}}$$

The rms values obtained (Table 2) demonstrate a good fit between the synthetic and observed ground velocities, although for the 2009 earthquake they are somewhat larger. The best overall fits correspond to the 2007 shock, and the Z component has the best fit for the two earthquakes.

To continue testing the velocity model, we also analysed the differences between the theoretical and observed P arrival times. The earthquakes origin time is taken from IGN, and it is well known. In Fig. 4a

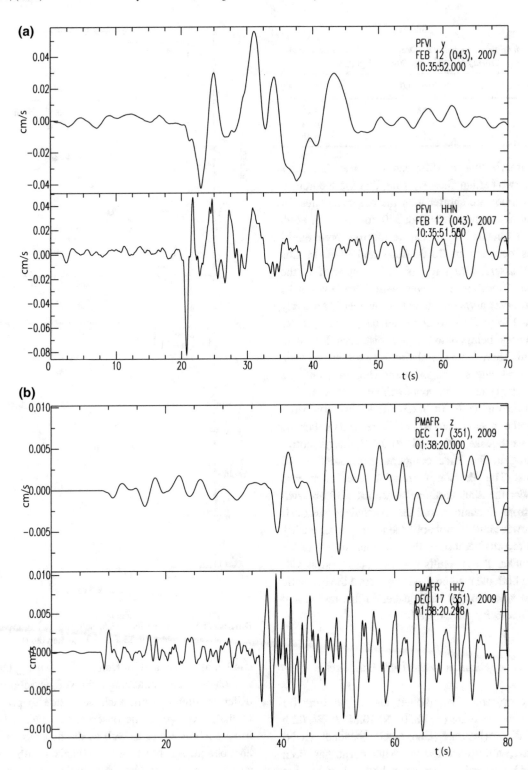

Figure 3

a synthetic (top) and observed (bottom) ground velocity (cm/s) for the NS component at PFVI station for the 2007 earthquake; **b** synthetic (top) and observed (bottom) ground velocity (cm/s) for the vertical component at PMAFR station for 2009 earthquake

Table 2

rms of the differences between observed and synthetics seismo-grams for 2007 and 2009 earthquakes

Component	rms 2007	rms 2009
EW	0.13	0.41
NS	0.17	0.53
Z	0.06	0.25

one observes that the differences are less than 1.5 s for the two earthquakes, showing that the theoretical arrival times are shorter than the observed ones for epicentral distances less than 350 km. Grandin et al. (2007a) obtained a similar result for three earthquakes occurred in 2003, 2004 and 2006 in this region, asserting that it was due to an error in the epicentral location. However, with these new results one observes a systematic behaviour that is probably related to the 3D crustal model used, with greater differences being obtained for distances less than 350 km. There is no correlation between arrival time differences and station azimuths (Fig. 4b), but this could simply reflect the poor azimuthal coverage.

To determine the fit in amplitudes, we compared the synthetic and observed not filtered PGVs. For the 2007 earthquake, we used 30 PGV values corresponding to the three components recorded at 10 stations. The Pearson correlation coefficient was 0.77. For the 2009 earthquake, we used 36 observations from 12 stations, and that coefficient was 0.51. This lower value is probably due to the poorer quality of the records because of the lower magnitude of the earthquake. These results may be considered as validating both the method used to generate the synthetic ground velocities for earthquakes in this region and the 3D velocity model used.

3. *Application to the 1969 Earthquake*

The epicentral coordinates and origin time for 1969 earthquake are (35.9850° N, 10.8133° W, 02 h 40 m 32 s, (Instituto Geográfico Nacional (IGN) database, http://www.ign.es). This event has been studied by several authors using short and long period records from WWSSN stations at teleseismic (30°–90°) distances. The focal mechanism of this earthquake was originally determined by Lopez Arroyo

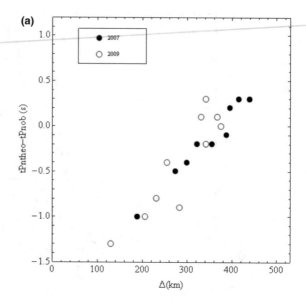

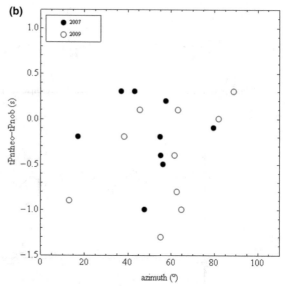

Figure 4
Differences between the theoretical and observed P arrival time versus epicentral distance (**a**) and azimuth (**b**)

and Udías (1972) and McKenzie (1972) using the first motion data, both obtaining a thrust fault solution but differing slightly from each other due to poor azimuthal coverage of the observations. Fukao (1973) using surface waves as well as the first motion data, also obtained a thrust fault mechanism with a small strike slip component. A similar solution was obtained by Buforn et al. (1988a) using P-wave first motion polarities. Grimison and Chen (1988) by inversion of P and SH waveforms obtained a similar focal mechanism, and explain that, for this large event, P and S waves were off-scale on most of the long period vertical components, but they reconstructed the vertical component at some stations. Finally, Grandin et al. (2007b) tested the focal mechanism solutions provided by these studies by comparing simulated and observed intensities, obtaining a similar solution. They also estimated fault dimensions and average slip. To summarize, one sees in Table 3 that these authors all agree that the mechanism corresponds to a thrust fault and obtain a similar orientation, but that there are relatively large differences in the estimated seismic moment and some differences in the focal depth.

To study the 1969 earthquake using the aforementioned observed data (Lisbon bridge accelerograph and intensity distribution) and the synthetic ground velocities, we first calculated the focal depth and seismic moment by comparing the synthetic and observed acceleration at 25th April Bridge (Lisbon, Portugal). The synthetic acceleration was obtained by differentiating the synthetic velocity that was estimated previously with the mentioned 3D velocity model and a fixed focal mechanism. The focal mechanism solution chosen was that proposed in Buforn et al. (1988a) and we considered at this step a point source in order to obtain the focal depth, with a Gaussian source time function. Different seismic moment from Table 3 and focal depth values between 20 and 40 km were tested, calculating the rms between observed and synthetic accelerations for each component. The best solution corresponded to a depth of 25 km and a seismic moment of 6.4×10^{20} N m (Mw = 7.8).

It was difficult to process the records at the Lisbon bridge and to make a comparison with the synthetic

values because of the different frequencies. With respect to the synthetic values, as was mentioned above, the velocity model reproduces better the low values (0.1–0.5 Hz) but those observed contain higher frequencies (0.2–2.5 Hz). We tested several filters, with the best results corresponding to a Butterworth band-pass between 0.2 and 0.5 Hz. While a perfect fit was not possible to find, by varying the focal depth and fixing the other focal parameters we were able to refine the depth on the basis of the rms difference between the synthetic and observed accelerations.

Figure 5 shows the results corresponding to the three components together with a picture of the Lisbon bridge. One notes that the observed records present higher frequencies even though the filter has been applied. The observed and synthetic PGA values are similar for the three components, with the maximum observed PGA corresponding to the EW component and being similar to the synthetic one. With regards to the frequencies, the poorest result corresponds to the NS component.

With these values for depth and seismic moment, and the focal mechanism of Buforn et al. (1988a) (231°/47°/126°), we generated synthetic ground velocities for 159 sites located in the Iberian Peninsula and northern region of Morocco where we have observed intensity values (López-Sánchez et al. 2019). To estimate the ground velocities, we used the E3D code (Larsen and Schultz 1995) and assumed as source model a finite fault with width equal to 20 km. The slip was assumed to be constant, and we took a Gaussian source time function. Several values of the fault length and different positions of the nucleation relative to the fault centre were tested, with a fixed rupture velocity equal to 2.5 km/s. The fault centre is

Table 3

Source parameters for 1969 earthquake obtained by different authors

	h (km)	Mo ($\times 10^{20}$ N m)	Mw	L (km)	Strike (°)	Dip (°)	Rake (°)
Lopez Arroyo and Udías (1972)	22	0.63	7.1	90	228	46	152
McKenzie (1972)					220	50	
Fukao (1973)	33	6.0	7.8	80	235	52	
Buforn et al. (1988a)	22	0.63	7.1	85	231	47	126
Buforn et al. (1988b)	22	3.6	7.7	82.5	231	47	126
Grimison and Chen (1988)	32	8.0	7.8	–	250	44	113
Grandin et al. (2007b)	25	6.0	7.8	82.5	233	49.5	116.5

located in the hypocentre location at a depth of 25 km. The preferred solution corresponds to a fracture length equal to 70 km. This is shorter than the values obtained by other authors (Table 3) who used different methods. We also tested several positions for the focus along the fault direction. Although the results did not change much, the solution that presented the best PGV distribution corresponded to a fault centre location.

As an example, Fig. 6a–c show the synthetic ground velocities for the three components at three places (location is shown in Fig. 7) with different epicentral distance and azimuths.

Figure 7 shows the geographical distribution of the PGV values in cm/s corresponding to the maximum horizontal component synthetic ground velocity. One observes that the highest values (of 10–12 cm/s) are located in the south of Portugal, and the lowest in the northern part of Morocco.

Figure 6

Synthetic seismograms for 1969 earthquake in cm/s at Vila do Bispo (Portugal, EMS98= VIII), Bairro (Portugal, EMS98= VII) and Souk el Arba des Beni Hassane (Morocco, EMS98= VI). For each place the epicentral distance and azimuth are shown. The seismograms correspond to **a** NS, **b** EW and **c** vertical components. The geographical location of these places are shown in Fig. 7

4. Instrumental Intensity

From PGV values, one can obtain the so called "instrumental intensity" I_{MM} using an empirical relationship such as those proposed for Western United States (Wald et al. 1999), Taiwan (Wu et al. 2003), Eastern North America (Kaka and Atkinson 2004), and Italy (Faenza and Michelini 2010). Since there is no specific relationship for SW Iberia because of the lack of earthquakes with M > 6 in this area, we use the aforementioned relationship, even though the

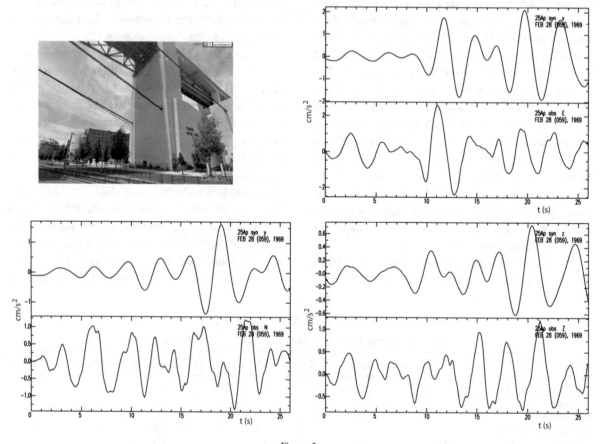

Figure 5

Picture of accelerometer location in the 25th April Bridge (Lisbon). Synthetic (top) and observed (bottom) acceleration for EW, NS and vertical components filtered with a Butterworth band-pass between 0.2 and 0.5 Hz

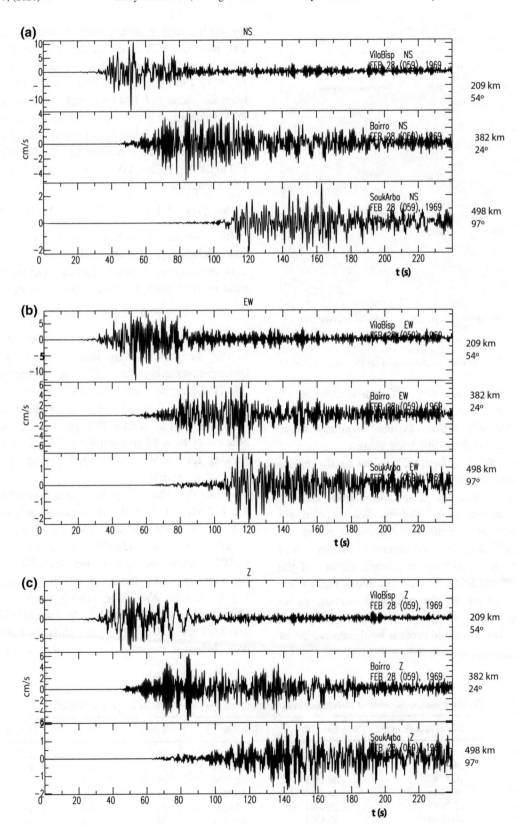

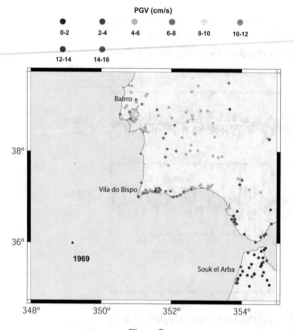

Figure 7
Synthetic PGV values (cm/s) distribution for the 1969 earthquake

tectonic features are very different. Grandin et al. (2007b) determined a regression fit to the I_{MM}-PGV values for this region, in the frequency range 0.1–0.5 Hz, but from simulated values.

To estimate I_{MM} from PGV values for the region, we deduce an empirical relationship between the PGV values recorded at seismic stations and the EMS98 intensities felt nearby. We selected six earthquakes that occurred in the region during the period 1990–2018 that had observed intensity values and that were registered at seismic stations of the IGN, WM and IPMA networks (Table 4, Fig. 8). We used a total of 41 observed PGV values, corresponding to the maximum horizontal component obtained from observed records by deconvolution of

the instrumental response without filtering, and we also used the PGV value at the 25th April Bridge from integrating the acceleration record of the 28 February 1969 earthquake. This site has the highest intensity value (VI EMS-98) and corresponds to a PGV value equal to 6.5 cm/s. The data were grouped into classes at 0.5 intensity intervals, and a linear fit was made between the intensity and the decadic (base 10) logarithm of the PGV values in cm/s (Fig. 9a, b). The result was:

$$I_{EMS98} = (4.8 \pm 0.1) + (1.7 \pm 0.1)\log(PGV),$$

with a correlation coefficient equal to 0.97. Faenza and Michelini (2010) obtained a similar result for Italy by using the Mercalli–Cancani–Sieberg (MCS) macroseismic scale, but with a greater slope:

$$I_{MCS} = 4.79 + 1.94 \log(PGV)$$

The accuracy of the empirical relation calculated in this study is strongly limited by the lack of instrumental earthquakes with intensities greater than VI. However, the low standard deviation for the coefficients ($\pm$ 0.1) and the high correlation coefficient (0.97) imply that the fit is good enough for this relation to be used to estimate the level of ground shaking for I < VI or to predict ground velocities from intensity data.

Since the relationship we obtained is only valid for I < VI, we used the expressions proposed by Wald et al. (1999) ($I_{MM} = 3.40 + 2.10 \log(PGV)$), Grandin et al. (2007b) ($I_{MM} = 3.50 + 3.50 \log(PGV)$), and Faenza and Michelini (2010) for all the intensity values and a combination of them: for $I \leq V$ we used Wald et al. (1999), for $V < I < VII$ Grandin et al. (2007b), and for $I \geq 7$ we took Faenza and Michelini (2010). The rms values for the difference between the observed and the instrumental

Table 4

Focal parameters for selected earthquakes occurred in St Vincent Cape region in the period 1990–2018

Date	Origin time (h m s)	Lat (°N)	Lon (°E)	h (km)	M_W	I_{max} (EMS-98)
29/07/2003	05:31:32	35.8067	− 10.6137	30	5.3	IV
13/12/2004	14:16:12	36.2654	− 9.9934	53	4.9	III
10/01/2006	10:57:40	36.1535	− 7.7136	46	5.1	II
12/02/2007	10:35:24	35.9100	− 10.4684	65	6.1	V
11/01/2008	00:21:47	36.4761	− 9.9978	56	4.4	IV–V
17/12/2009	01:37:49	36.4702	− 10.0318	56	5.5	V

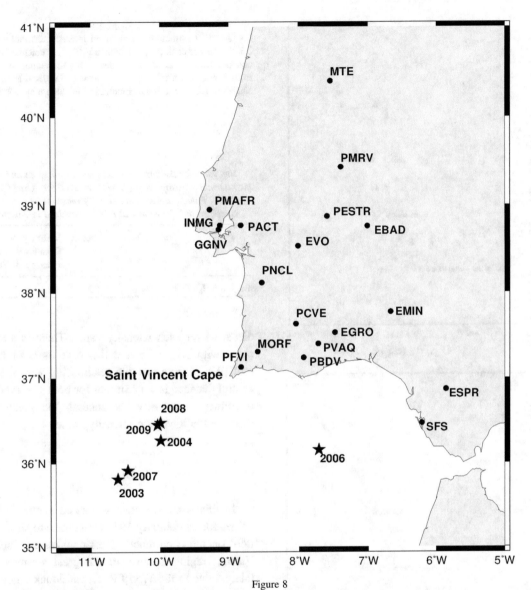

Figure 8
Epicentral location of the selected earthquakes with observed intensity value, and the broad band seismic station used to calculate the empirical relationship between EMS98 intensity and observed PGV

intensity showed the best fit to be that of the combination of the three expressions (Table 5).

Figure 10 shows the corresponding maps of the geographical distribution of the observed intensities compared with the intensities predicted by the cited empirical relations. That of Wald et al. (1999) does not reproduce the high observed EMS98 intensity values located at SW of Portugal. The same is the case with that of Grandin et al. (2007b) which over-estimates intensities for the other places. That of Faenza and Michelini (2010) reproduces well the

high intensity but shows greater intensities than those observed for other values. Finally, the best-fitting proposed combined relationship has the intensities being highest along the southwestern coast of Portugal and Spain and decreasing inland. Some anomalous high values close to Lisbon are also well reproduced.

In Fig. 11, the observed intensity and the I_{MM} (in red) obtained using the combined relation are represented versus the epicentral distances. The error bars show the standard deviation for the epicentral

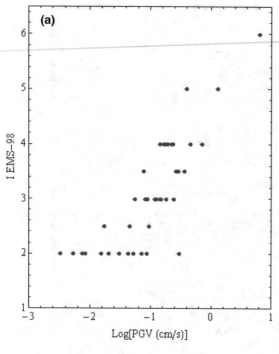

▶

Figure 10

For the 1969 earthquake: **a** observed intensity distribution from López-Sánchez et al. (López-Sánchez 2019). Instrumental intensity obtained from synthetic PGV values using the empirical relation from **b** Wald et al. (1999), **c** Grandin et al. 2007b, **d** Faenza and Michelini (2010) and **e** the combination of these relations used in this work

Table 5

rms value for the difference between the observed and the instrumental intensity using the Wald et al (1999), Grandin et al. (2007b), Faenza and Michelini (2010) relations as well as the combined and the empirical relation obtained in this study

	Wald	Grandin	Faenza	Wald $I \leq V$ Grandin $V < I < VII$ Faenza $I \geq VII$
rms $I_{ob} - I_{MM}$	1.2	1.0	1.2	0.7

distances for each intensity value. There is a single place with $I_{MM} = 3.5$ and therefore its error bar is null. We observe that the relationship with the epicentral distance is very similar for both observed and instrumental intensity. In general, the greater the distance the lower the intensity value.

5. Discussion

In this work we have generated synthetic PGV values for 28 February 1969 earthquake to shed some light on the occurrence of earthquakes in Cape St. Vincent region because of the great seismological interest due to the 1755 Lisbon earthquake. For that, we have used the accelerograph record at Lisbon and the observed intensities in the Iberian Peninsula and northern part of Morocco.

We compared the three components acceleration record of the 1969 earthquake with the synthetic accelerations. Although the synthetic and observed accelerations present different frequencies (Fig. 5), one notes that the PGA values are similar, which implies that the 3D velocity model and the method used are appropriate for generating the PGV values. In addition, we obtained a focal depth equal to 25 km and a seismic moment of 6.4×10^{20} Nm.

We generated the synthetic PGV values at regional distances and the simulated intensities using

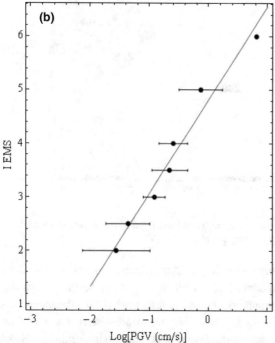

Figure 9

a observed intensity values (EMS98 scale) versus log(PGV) from earthquakes occurred in the region (Table 4); **b** observed intensity value with the standard deviation bar for log(PGV) together the obtained regression line

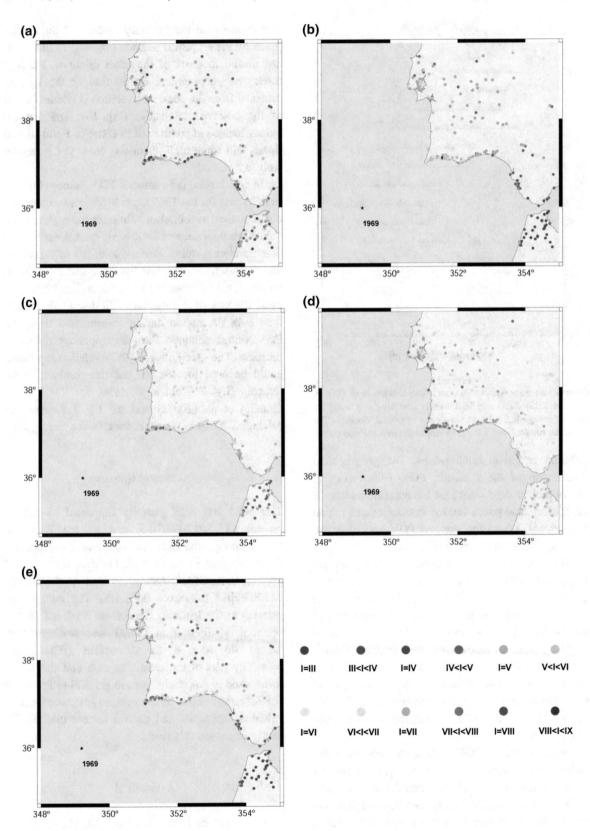

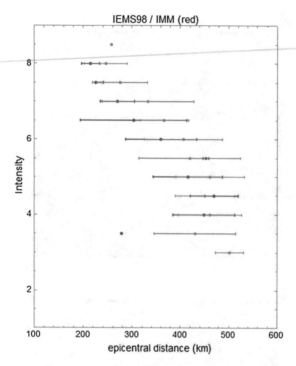

Figure 11

Observed intensity distribution from López-Sánchez et al. (López-Sánchez 2019) (blue) and instrumental intensity (red) using the combined empirical relation versus the epicentral distance. The deviation standard bars for the epicentral distance are also shown

different empirical relationships. The geographical distribution of the synthetic PGV values (Fig. 7) shows a clear dependence on the epicentral distance except for some points located in the southern region of Portugal, where the greatest PGV and observed intensity values are located (Fig. 10a).

We also obtained an empirical relation between I_{EMS98} and observed PGV values from the earthquakes that occurred in SW Iberia. There is no specific law for this zone, and it is fundamental in seismic hazard studies to estimate the ground shaking resulting from an earthquake. We obtained a sufficiently good fit for this relation to be used to estimate the level of ground shaking or to predict ground velocities from intensity data in the region for $I < VI$. In the future, it could be improved with more observations.

From synthetic PGV values, we obtained the instrumental intensity for 1969 earthquake using the most commonly used relations and the law obtained in this present study. Taking into account the rms values, the most suitable relation is a combination

that depends on the intensity value, and the results given by the empirical relation obtained in this work are similar to those of the other relations. Furthermore, the geographical distribution of the I_{MM} as obtained from the combined relation is similar to that of the observed intensities (Fig. 10). The greatest values located at southwestern coast of Portugal and Spain, and some high intensities close to Lisbon are well reproduced.

In conclusion, the synthetic PGV values obtained in this study for the 1969 Cape St Vincent earthquake could be used as reference values for the region. Such values are fundamental for seismic hazard studies and seismic alert systems. Furthermore, the methodological approach would allow the PGV and intensity to be simulated for other events in the region. Bearing in mind the lack of earthquakes with magnitude greater than 6 in the region during instrumental time, the PGV values obtained form an important reference database. The specific I_{EMS98}-PGV relationship found could be used for $I < VI$, but the combined relationship ($I \leq V$ Wald et al. 1999, for $V < I < VII$ Grandin et al. (2007a) and for $I \geq 7$ Faenza and Michelini 2010) is the most suitable for the region.

Acknowledgments

This work has been partially supported by MEIC, project CGL2017-86070-R and by the European Union through the European Regional Development Fund, included in the COMPETE 2020 through ICT project (UID/GEO/04683/2019) and SFRH/BSAB/143063/2018 (reference 0061205). The authors are grateful to the Instituto Geográfico Nacional (IGN), Western Mediterranean (WM) and Instituto Português do Mar e da Atmosfera (IPMA) for providing part of the catalogue data and the waveforms used in this study. We are grateful to Professor Jiri Zahradnik and to an anonymous reviewer for their valuable comments and careful review that significantly improved this work.

Appendix A

See appendix Figs. 12, 13, 14, 15, 16, 17.

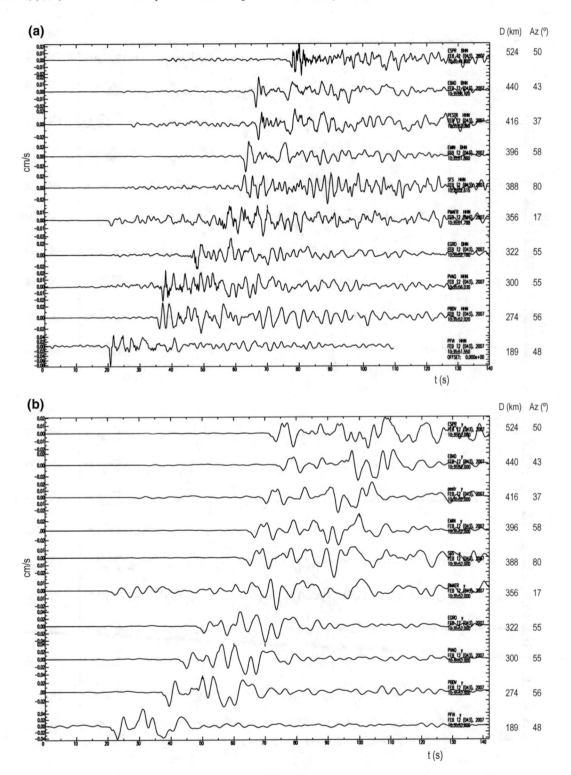

Figure 12
Ground velocity (cm/s) for 2007 earthquake NS component: **a** observed **b** synthetic. The epicentral distance and azimuth are shown for each station

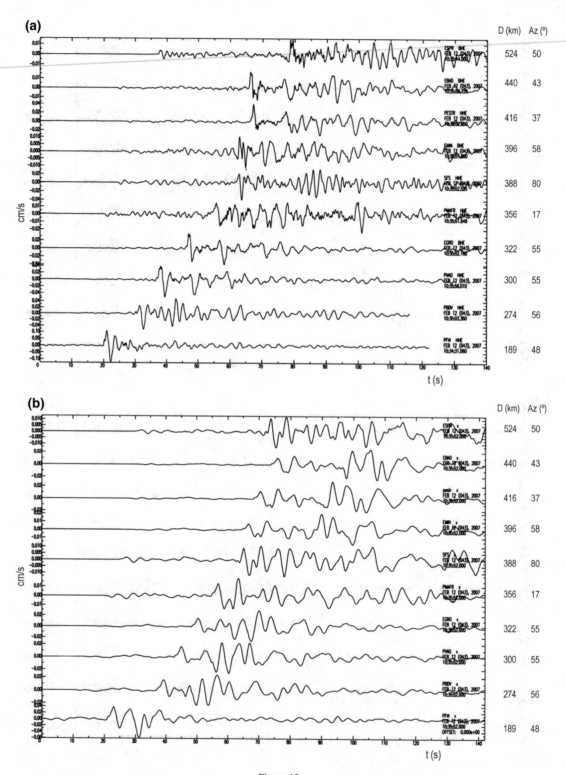

Figure 13
Ground velocity (cm/s) for 2007 earthquake EW component: **a** observed, **b** synthetic. The epicentral distance and azimuth are shown for each station

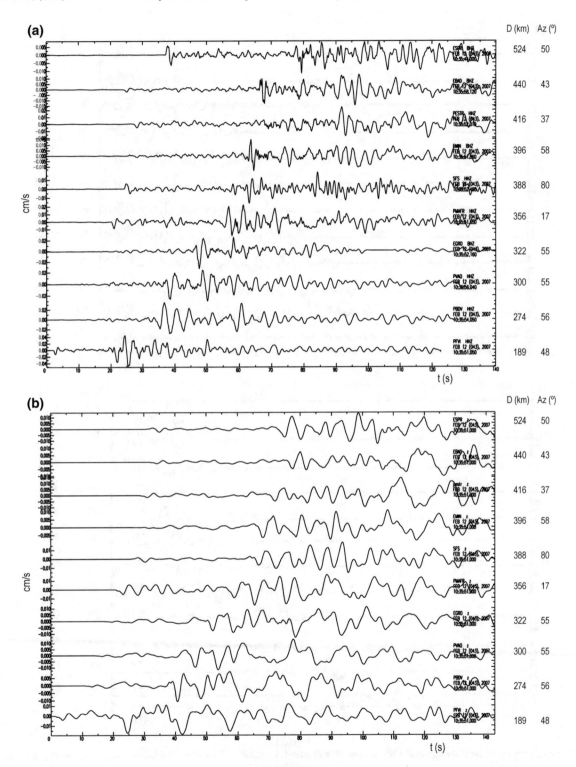

Figure 14
Ground velocity (cm/s) for 2007 earthquake Z component: **a** observed, **b** synthetic. The epicentral distance and azimuth are shown for each station

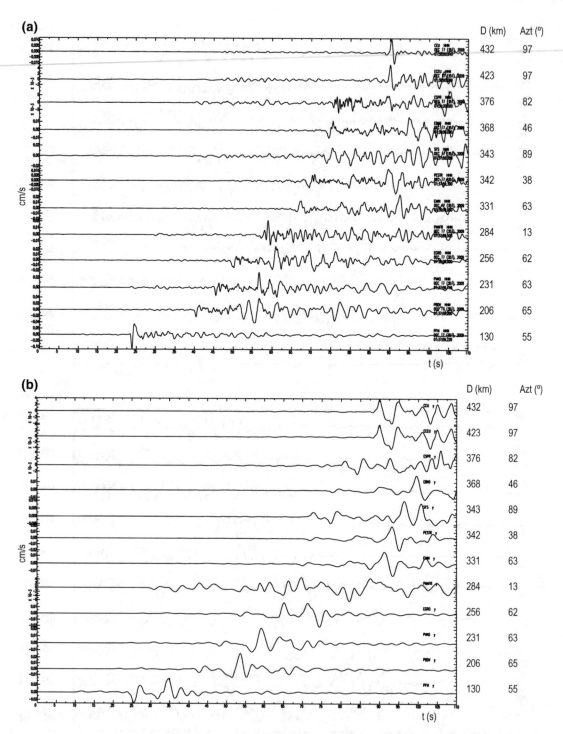

Figure 15
Ground velocity (cm/s) for 2009 earthquake NS component: **a** observed, **b** synthetic. The epicentral distance and azimuth are shown for each station

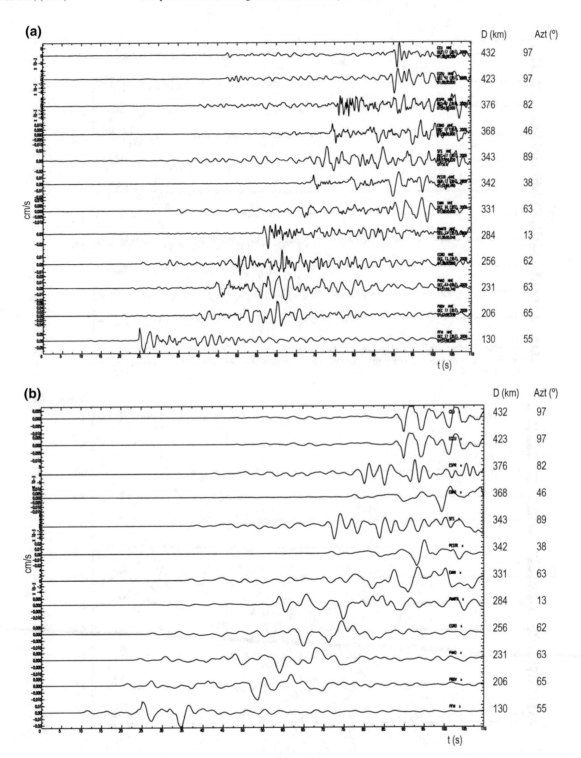

Figure 16

Ground velocity (cm/s) for 2009 earthquake EW component: **a** observed, **b** synthetic. The epicentral distance and azimuth are shown for each station

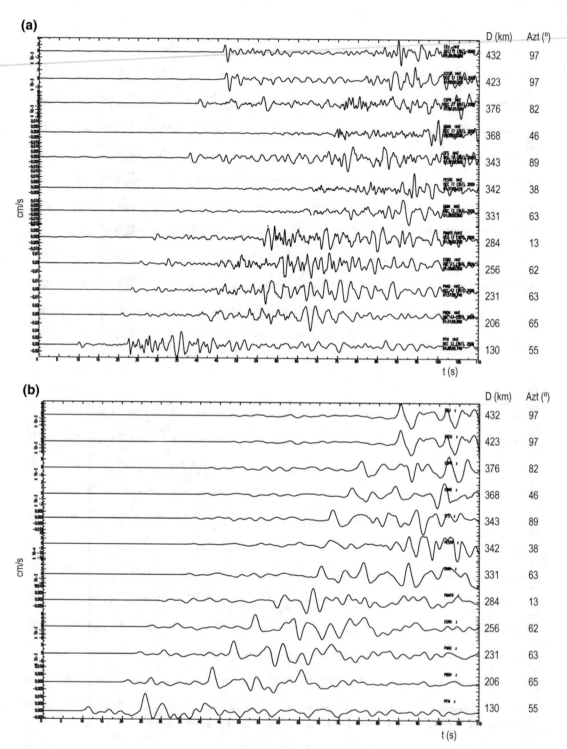

Figure 17

Ground velocity (cm/s) for 2009 earthquake Z component: **a** observed, **b** synthetic. The epicentral distance and azimuth are shown for each station

Publisher's Note   Springer Nature remains neutral with regard to jurisdictional claims in published maps and institutional affiliations.

REFERENCES

Baptista, M. A., Miranda, J. M., Chiericci, F., & Zitellini, N. (2003). New study of the 1755 earthquake source based on multichannel seismic survey data and tsunami modeling. *Natural Hazards and Earth Systems Sciences, 3,* 333–340.

Baptista, M. A., Miranda, J. M., & Luís, J. F. (2006). In search of the 31 March 1761 earthquake and tsunami source. *Bulletin of the Seismological Society of America, 96*(2), 713–721. https://doi.org/10.1785/0120050111.

Buforn, E., Udías, A., & Colombás, M. A. (1988a). Seismicity, source mechanisms and tectonics of the Azores-Gibraltar plate boundary. *Tectonophysics, 152,* 89–118.

Buforn, E., Udías, A., & Mézcua, J. (1988b). Seismicity and focal mechanisms in south Spain. *Bulletin of the Seismological Society of America, 78,* 2008–2224.

Cunha, T. A., Matias, L. M., Terrinha, P., Negredo, A. M., Rosas, F., Fernandes, R. M. S., et al. (2012). Neotectonics of the SW Iberia margin, Gulf of Cadiz and Alboran Sea: a reassessment including recent structural, seismic and geodetic data. *Geophysical Journal International, 188,* 850–872. https://doi.org/10.1111/j.1365-246X.2011.05328.x.

Faenza, L., & Michelini, A. (2010). Regression analysis of MCS intensity and ground motion parameters in Italy and its application in ShakeMap. *Geophysical Journal International, 180,* 1138–1152. https://doi.org/10.1111/j.1365-246X.2009.04467.x.

Fukao, Y. (1973). Thrust faulting at a lithospheric plate boundary: the Portugal earthquake of 1969. *Earth and Planetary Science Letters, 18,* 205–216.

Grandin, R., Borges, J. F., Bezzeghoud, M., Caldeira, B., & Carrilho, F. (2007a). Simulations of strong ground motion in SW Iberia for the 1969 February 28 (Ms = 8.0) and the 1755 November 1 (M ∼ 8.5) earthquakes–I. Velocity model. *Geophysical Journal International, 171,* 1144–1161.

Grandin, R., Borges, J. F., Bezzeghoud, M., Caldeira, B., & Carrilho, F. (2007b). Simulations of strong ground motion in SW Iberia for the 1969 February 28 (Ms = 8.0) and the 1755 November 1 (M ∼ 8.5) earthquakes–II. Strong ground motion simulations. *Geophysical Journal International, 171,* 807–822.

Grimison, N., & Chen, W. (1988). Focal mechanisms of four recent earthquakes along the Azores-Gibraltar plate boundary. *Geophysical Journal of the Royal Astronomical Society, 92,* 391–401.

Gutscher, M. A., Baptista, M. A., & Miranda, J. M. (2006). The Gibraltar Arc seismogenic zone. Part 2: Constraints on a shallow east dipping fault plane source for the 1755 Lisbon earthquake

provided by tsunami modeling and seismic intensity. *Tectonophysics, 426*(1–2), 153–166.

Kaka, S. I., & Atkinson, G. M. (2004). Relationships between instrumental ground-motion parameters and Modified Mercalli Intensity in Eastern North America. *Bulletin of the Seismological Society of America, 94,* 1728–1736.

Larsen, S.C. & C.A. Schultz (1995). ELAS3D, 2D/3D elastic finite-difference wave propagation code, Lawrence Livermore National Laboratory, UCRLMA-121792, 18 pp

Levander, A. R. (1988). Fourth-order finite-difference *P-SV* seismograms. *Geophysics, 53,* 1425–1436.

Lopez Arroyo, A., & Udías, A. (1972). Aftershock sequence and focal parameters of the February 28, 1969, earthquake of Azores–Gibraltar fracture zone. *Bulletin of the Seismological Society of America, 62,* 699–720.

López-Sánchez, C., Lozano, L., Buforn, E., Martínez-Solares, J. M., Cantavella, J. V., & Udías, A. (2019). Re-evaluation of seismic intensity for the February 28, 1969 main shock and relocation of main aftershocks. In *Proceeding of the Workshop "Earthquakes and tsunamis in Iberia: 50th years of the 1969 Saint Vincent earthquake (M = 8.0)".* Madrid, Spain: Universidad Complutense de Madrid.

McKenzie, D. P. (1972). Active tectonics of the Mediterranean region. *Geophysical Journal of the Royal Astronomical Societ, 30,* 109–185.

Pro, C., Buforn, E., Bezzeghoud, M., & Udías, A. (2013). The earthquakes of 29 July 2003, 12 February 2007, and 17 December 2009 in the region of Cape Saint Vincent (SW Iberia) and their relation with the 1755 Lisbon earthquake. *Tectonophysics, 583,* 17–27. https://doi.org/10.1016/j.tecto.2012.10.010.

Stich, D., Ancilla, F., Pondrelli, S., & Morales, J. (2007). Source analysis of the February 12th 2007, Mw 6.0 Horseshoe earthquake: Implications for the 1755 Lisbon earthquake. *Geophysical Research Letters, 34,* L12308. https://doi.org/10.1029/2007gl030012.

Udías, A., López Arroyo, A., & Mézcua, J. (1976). Seismotectonics of the Azores–Alboran region. *Tectonophysics, 31*(1976), 259–289.

Vilanova, S. P., Nunes, C. F., & Fonseca, J. F. B. D. (2003). Lisbon 1755: A case of triggered onshore rupture? *Bulletin of the Seismological Society of America, 93,* 2056–2068.

Wald, D. J., Quitoriano, V., Heaton, T. H., Kanamori, H., Scrivner, C. W., & Worden, C. B. (1999). TriNet "ShakeMaps": rapid generation of instrumental ground motion and intensity maps for earthquakes in southern California. *Earthquake Spectra, 15,* 537–556.

Wu, Y. M., Teng, T. L., Shin, T. C., & Hsiao, N. C. (2003). Relationship between peak ground acceleration, peak ground velocity, and intensity in Taiwan. *Bulletin of the Seismological Society of America, 93,* 386–396.

Zitellini, N. L. A., Mendes, D., Córdoba, J., Dañobeitia, R., Nicolich, G., Pellis, A., et al. (2001). Source of 1755 lisbon earthquake and tsunami investigated. *EOS Transactions of the American Geophysical Union, 82*(26), 290–291.

(Received July 8, 2019, revised December 16, 2019, accepted December 17, 2019, Published online January 15, 2020)

Pure Appl. Geophys. 177 (2020), 1831–1844
© 2019 Springer Nature Switzerland AG
https://doi.org/10.1007/s00024-019-02329-7

Human Losses and Damage Expected in Future Earthquakes on Faial Island–Azores

J. Fontiela,[1] ⓘ P. Rosset,[2] ⓘ M. Wyss,[2] ⓘ M. Bezzeghoud,[1] ⓘ J. Borges,[1] ⓘ and F. Cota Rodrigues[3] ⓘ

Abstract—Since the 15th century, the death toll due to large earthquakes has reached approximately 6300 in the Azores and 17 for Faial Island. The likely number of fatalities and injuries in future large earthquakes (M6+) in Faial Island is estimated using the software QLARM (Quake Loss Assessment for Response and Mitigation) and its dataset, validated at the regional scale. The current population for the 13 settlements on this island is extrapolated from the 2001 and 2011 census. The distribution of buildings into EMS-98 vulnerability classes is based on detailed census information and damage reports written after the 1980 (M7.2) and 1998 (M6.2) earthquakes. The most appropriate ground motion prediction equation (i.e. Shebalin in Regularities of the natural disasters (in Russian), Nauki o zemle, Znanie, vol 11, p 48, 1985) is selected and coupled with site amplification information to adjust calculated ground shaking with observed intensities for the aforementioned events. The good agreement of fatalities and injuries calculated by QLARM with the observed numbers in both earthquakes was the motivation for proposing two scenarios for likely future earthquakes of M6 and 6.9, offshore and inland, respectively. Based on these scenarios, fatalities, and injuries may range between 110–620 and 330–1750, respectively, depending on the likely large earthquake.

Key words: Earthquake loss-estimation, QLARM, attenuation models, EMS-98, volcanic islands, Azores.

1. Introduction

In the past, the population of the Azores islands has suffered several large earthquakes, suggesting

Electronic supplementary material The online version of this article (https://doi.org/10.1007/s00024-019-02329-7) contains supplementary material, which is available to authorized users.

[1] Physics Department of the School of Technology and Science, Institute of Earth Sciences (ICT), Universidade de Évora, Colégio Luís António Verney, Rua Romão Ramalho, 59, 7000-671 Évora, Portugal. E-mail: jfontiela@uevora.pt
[2] International Centre for Earth Simulation Foundation (ICES), Geneva, Switzerland.
[3] Institute of Agricultural and Environmental Research and Technology of the Azores, Faculdade de Ciências Agrárias e do Ambiente, Universidade dos Açores, Angra do Heroísmo, Portugal.

that a worst-case scenario should be considered for risk preparedness and prevention. Wyss (2017a) showed that it is possible to reduce the number of losses if the stakeholders use loss estimations to engage necessary retrofitting and evacuation measures. Loss estimation depends on reliable models, which should be able to predict fatalities within a factor of two or three of the observed values. In the specific case of the Azores region, the rescue operations after a strong earthquake could be difficult, due to the intervention delay of the rescue teams amongst the different islands.

The software tool QLARM is used to estimate building damage and losses as it has been used for different regions of the world such as North India (Wyss et al. 2017), Central Myanmar (Wyss 2008), Mexico (Wyss and Zuñiga 2016), Himalayas (Wyss 2005, and Wyss et al. 2018), Algeria (Rosset and Wyss 2017), Southern Sumatra and Central Chile (Wyss 2010).

The present paper reports on an effort to build an accurate dataset to be used in QLARM in the volcanic islands of the Azores. It includes a detailed compilation of the building stock at the scale of settlements, the population distribution, as well as the soil conditions which may amplify seismic waves. Efforts have also been made to validate both the ground motion and vulnerability models using past observations from several damaging earthquakes.

The seismicity of the Azores is first discussed, pointing out the specific tectonic and geological context combining seismic and volcanic activities. Secondly, the dataset used in QLARM is explained and validated using past observations to run scenarios of likely earthquakes for Faial Island, based on knowledge of regional tectonics.

2. Seismicity in the Azores

The Azores archipelago is a seismically active zone, as attested by historical and instrumental earthquakes plotted in Fig. 1. Since the 15th century, several large earthquakes have caused heavy damage with Modified Mercalli Intensities (MMI) ranging from VII to XI, killing more than 6300 people (Nunes et al. 2001), as listed in Table 1. The most severe earthquakes that have occurred in the Central Group affected Terceira Island in 1614 and 1980, São Jorge Island in 1757, and Faial Island in 1926, 1958 and 1998. The causes of the 1980 and 1998 earthquakes were discussed by several authors (Hirn et al. 1980; Grimison and Chen 1986, 1988; Buforn et al. 1988; Borges et al. 2007). Following the 1998 earthquake, studies of the seismic sequence (Matias et al. 2007), and the crustal structure (Dias et al. 2007) were published.

The May 24, 1614 earthquake affected mainly the eastern part of Terceira Island with heavy damage to the building stock (maximum MMI of X) and killed more than 200 people. The July 9, 1757 earthquake (M7.4), the second deadliest earthquake in the Azores, had a maximum MMI of XI and a death toll of 1046 persons (Machado 1949). The earthquake of August 31, 1926, at 10h40 with mb 5.3-5.9 (Nunes et al. 2001) caused nine fatalities and more than 200 injuries (Lima 1934). According to Agostinho (1927), 15% of the buildings collapsed in the city of Horta, the area with the highest intensity (MMI = X), as shown in Fig. 2. September 27th, 1957 started a volcanic eruption off the west coast of Faial Island and lasted until October 1958. A swarm of around 450 earthquakes were felt during the night of May 12, 1958, two of them with a maximum MMI of X (Machado et al. 1962). Lobão (1999) counted 273 collapsed houses among the 508 reported damaged, but no fatalities were registered.

On the 1st of January 1980, an M_S7.2 earthquake occurred in the evening affecting Terceira, São Jorge, and Graciosa Islands. The death toll reached 61 with more than 300 injured and around 5400 houses collapsed or heavily damaged (MMI = IX), leaving

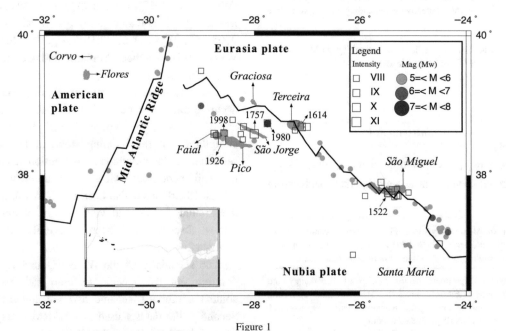

Figure 1

Seismicity map of the Azores archipelago. Colored circles instrumentally located seismicity with magnitude M ≥ 5 for the period 1926–2017 and empty blue squares show historical earthquakes with MMI ≥ VIII since 1522 (Sources: Nunes et al. 2004; Fontiela et al. 2017). The year of the historical event is indicated when more than four deaths were reported (Table 1). The islands are filled in light grey and the main faults forming the boundaries of the three tectonic plates are drawn in black (source: Bird 2003)

Table 1

Main characteristics of the earthquakes for which more than four fatalities were reported. These earthquakes are plotted in the map of Fig. 1

Date and time (local time) (yyyy/mm/dd hh:mm)[d]	Long (°E)	Lat (°N)	Magnitude	Maximum intensity	Island	Death toll
1522/10/22 ~ 02: –	− 25.400	37.400		X[b]	São Miguel	~ 5000
1614/05/24 15:15	− 27.100	38.700		X[b]	Terceira	~ 200
1757/07/09 –:–	− 28.000	38.600	M 7.4[a]	XI[b]	São Jorge	1043
1926/08/31 10:40	− 28.600	38.500	Mb 5.3	X[b]	Faial	9
1980/01/01 16:42	− 27.605	38.590	M_S 7.2	VIII/IX[b]	Terceira	61
1998/07/09 05:19	− 28.523	38.634	M_W 6.2	VIII[c]	Faial	8

[a]Magnitude estimated by Machado (1949)

[b]IMM

[c]EMS98

[d]Local time = GMT − 1

Figure 2
Illustration of a collapsed house in Flamengos during the 1926 earthquake in Faial Island (courtesy of Paulo Borges)

21,296 people homeless. The event caused heavy damage in the city of Angra do Heroísmo, as well as in other settlements of Terceira Island. In São Jorge Island, it caused heavy damage to the settlements around the epicenter. In Graciosa Island, the building stock also suffered considerable damage.

On the 9th of July 1998, an Mw 6.2 earthquake occurred at 05h19 and was largely felt in Faial but also in Pico, São Jorge, Graciosa, Terceira and São Miguel Islands (Senos et al. 1999; Fernandes et al. 2002; Matias et al. 2007; Borges et al. 2007). At the end of October 1998, 10,600 aftershocks were recorded, and the number rose to around 15,000 by the end of 2003. The population felt 410 events in the first four months (Matias et al. 2007). Despite the magnitude and the early time of its occurrence, only eight fatalities (mainly children and seniors), and 128 hospitalized people were reported (Gonçalves 2008). According to Ferreira and Oliveira (2008), around 35% of the building stock was affected, the strongest damage (degrees 4 and 5 on a scale from 0, no damage, to 5, collapse) occurring at epicentral distances up to 19 km. After the earthquake, several surveys were carried out to estimate the damage in the building stock. From these surveys, Ferreira (2008) and Senos et al. (2008) argue that the strongest damage occurred around the epicenter, with a maximum intensity of VIII; moreover, in the settlements of Flamengos and Castelo Branco (cf their location on Fig. 4) the damage was relatively high compared with neighbor settlements. In Flamengos, Pena et al. (2001) estimated that the soil amplification increases by two units the intensity that would be expected for hard rock sites, mainly due to the volcanic alluvium deposits, which have a strong frequency response around 9–10 Hz. In Castelo Branco, still, no clear explanation is given to justify the high level of damage.

3. Use of QLARM to Estimate Losses in the Azores

QLARM is a software tool that has provided human loss estimates for over 10 years for large earthquakes worldwide, in real-time (e.g. Wyss 2014), as well as scenario modeling, worldwide (e.g., Wyss et al. 2006, 2017, 2018; Wyss 2008, 2017a, b; Trendafiloski et al. 2009; Wyss and Wu 2014; Wyss and Zuñiga 2016). QLARM dataset includes around 1.93 million settlements, each documented with name, coordinates, building and population distribution, and in some cities, site conditions (Wyss 2010; Rosset et al. 2015). Using supplied magnitude, depth and epicenter coordinates, QLARM estimates ground motion at each settlement around the earthquake source. Some intensity attenuation models (IPE) and ground motion prediction equations (GMPEs) are available to adapt seismic attenuation to the tectonic context of the investigated region. When available, a soil condition factor is applied for a given settlement. The estimation of the expected damage to a building stock is performed using the european macroseismic method for vulnerability analysis, hereafter noted as

EMMVA (Trendafiloski et al. 2009). The method itself was developed by Lagomarsino and Giovinazzi (2006) using the implicit vulnerability model included in the European Macroseismic Scale (EMS-98 1998). The EMMVA uses seismic intensity to express the earthquake demand, which can be derived from the intensity, or any other ground motion parameter using appropriate IPE relationships.

The building stock can be modeled either by the EMS-98 vulnerability classes or corresponding building types. The concept of "vulnerability index" and its calibration makes QLARM flexible and gives the possibility to define specific building types not included in EMS-98, consequently enlarging the database of available building types and corresponding vulnerability models for worldwide use. The mean damage grade M_D is calculated from the damage rate divided into six damage degrees (from 0 to 5, where 0 stands for no damage and 5 for collapse). The EMMVA is extended with a casualty estimation module that uses the concept of an initial casualty matrix, hereafter ICM (Vacareanu et al. 2004; Trendafiloski 2007). The ICM is the probability of occurrence $P(D_i|C_i)$ of a certain casualty state C_i for the expected damage degree D_i.

When data are available, discrete city models have been developed consisting of the division of a settlement into a set of districts corresponding to the center of the administrative district. In this case, each district is documented, increasing the resolution and therefore the details in the loss calculation (see examples in Parvez and Rosset 2014).

As much as possible, a procedure of validation of QLARM using observed earthquake consequences is performed to adjust the different parameters involved in the loss calculation for a specific region. It includes, by order of importance, (1) choice of the attenuation model that best fits the observed ground shaking, (2) adjustments of the building and population distribution, (3) improvement of the collapse models and (4) of the casualty matrices. For the last three steps, a comparison is needed between past earthquake losses with calculated ones to verify the model. QLARM is considered validated when the difference between observed and calculated losses is less than a factor 2.

Table 2

Population and building distributions by settlements of Faial Island, as well as their distribution by EMS-98 classes are based on 2011 census data. The incremental value given to intensity is provided to take into account site amplification

Settlements	Population (2011)	Number of buildings	Amplification factor (in units of EMS-98)	Distribution of buildings by EMS-98 classes (%)				Distribution of population by EMS-98 classes (%)			
				A	B	C	D	A	B	C	D
Capelo	486	416		1.5	58.8	17.3	22.3	1.9	60.9	16.9	20.3
Castelo Branco	1304	563	+ 1	35.4	14.7	33.7	16.1	33.6	14.3	35.7	16.4
Cedros	906	547	+ 1	4.7	76.7	16.3	2.3	3.9	70.6	23.7	1.9
Feteira	1896	768	+ 1	7.9	53.9	32.9	5.3	7.5	52.9	34.0	5.6
Flamengos	1599	594	+ 2	1.6	40.5	55.3	2.6	1.6	40.3	55.6	2.6
Horta (Angústias)	2402	996		3.8	48.6	22.8	24.8	3.6	46.5	25.2	24.7
Horta (Conceição)	1138	459		8.3	41.4	21.4	28.8	6.6	36.8	23.0	33.6
Horta (Matriz)	2404	856		1.2	64.4	13.2	21.1	1.3	57.1	17.0	24.7
Pedro Miguel	753	313	+ 1	25.4	13.2	45.6	15.8	24.2	13.3	46.4	16.1
Praia do Almoxarife	834	361	+ 1	4.3	37.5	52.1	6.1	4.7	38.4	50.9	6.0
Praia do Norte	250	225		5.8	75.3	17.4	1.6	5.2	49.3	43.7	1.8
Ribeirinha	426	175	+ 1	1.3	23.5	73.8	1.3	1.3	18.9	78.4	1.4
Salão	401	174	+ 1	11.3	16.5	53.4	18.8	10.7	14.3	55.7	19.2

3.1. Validation of the Ground Motion Model

QLARM includes several published attenuation relationships that can be used in different tectonic contexts (Rosset et al. 2015). For example, intensities may be calculated using the equations by Shebalin (1968, 1985), Ambraseys (1985) or by downloading pre-calculated shakemaps in the USGS format (Allen et al. 2008). PGA relationships are those of Huo and Hu (1992), Ambraseys et al. (1996), Youngs et al. (1997), Munson and Thurber (1997), and Boore et al. (1997). These PGA relationships are coupled to the Wald et al. (1999) to convert PGA into intensity.

The earthquake catalog of Nunes et al. (2004) is used to compare with the calculated data from QLARM using different attenuation relationships for both 1980 and 1998 earthquakes. The 1998 Faial earthquake (M_W6.2) includes reports from 63 settlements located between 10 and 120 km of the epicenter with intensities ranging from III to VIII, and the 1980 Terceira earthquake (Ms7.2) includes reports from 56 settlements at distances from 25 km to 75 km with intensities between IV+ to VIII+. We chose these two earthquakes to validate IPE because are the ones with detailed data available and were subject of several studies. An estimate of the intensity's increase in several settlements compared to the surrounding ones is possible using the information provided by the macroseismic surveys and by

geological. For eight settlements of the Faial Island, an increment intensity of one or two units is attributed as shown in Table 2. Intensities are estimated using the Shebalin's (1968, 1985) formula expressed by:

$$I = C1 \times M - C2 \times \log\left(r^2 + h^2\right)^{0.5} + C3, \quad (1)$$

where I is the intensity (MMI scale), M the magnitude, r the hypocentral distance, h the depth, and C1, C2 and C3 are constants as described hereafter.

The performance of Eq. (1) has been compared to the one proposed by Ambraseys (1985) and Ambraseys et al. (1996) for the European context or by Munson and Thurber (1997) developed specifically for the volcanic Big Island in Hawaii. The last two, providing PGA values were associated with the Wald et al. (1999) relation to convert PGA into intensity. The choice of the Eq. (1) leads to better matches of calculations with observations compare to the other tested relations. The large scatter of intensity data for both 1980 and 1998 earthquakes at a given distance (up to 4 units of intensity at short distance) suggests that the default value of 1.5 used in QLARM for the slope of the curve (C1 factor of the Eq. 1) is the most pertinent one. For the 1998 event, intensity values decrease rapidly at a short distance (10–50 km) for settlements in Pico and Faial Islands, indicating a strong attenuation, which is simulated by tuning the value of the factor C2 of Eq. 1 which controls the attenuation relative to epicentral distance. The factor C3 is fixed to 3.5 which is a value generally used in QLARM (Wyss et al. 2018). In Fig. 3, the calculated MMI using Shebalin (1968, 1985) equation are compared with observed ones for both Terceira (1980) and Faial (1998) earthquakes. Figure 3a, b plot residuals between observed and calculated intensities (small dots in Fig. 3) for the 1980 and 1998 earthquakes, respectively. By adjusting the value of C2 to 5.2 and 5.7, the RMS error is minimized to 0.7 for the 1980 and 1998 earthquakes versus calculated ones, respectively. The residuals for data grouped by 5 km bins (black dots of Fig. 3) are lower than ± 0.5 intensity units on average in both cases. The obtained values are higher when using the Munson and Thurber (1997) relationship for both events. At this stage, one could consider that the ground motion can correctly estimate the variability of the intensity field for further earthquake scenarios. The variation of C2 value is a minor factor of uncertainty compare to uncertainties introduced by other factors in the risk analysis.

3.2. Construction Practices and Population on Faial Island

The elements at risk in QLARM concern both the buildings and their occupants. The level of resistance to the ground shaking of the buildings is expressed in terms of EMS-98 classes ranging from A to F, A being the least resistant buildings and F the most resistant buildings. The damage surveys conducted by Costa and Arêde (2006) after the 1998 earthquake in Faial, as well as the compilation of main characteristics of buildings done by Neves et al. (2012), are used to distribute the building stock of each settlement of the island into three main types:

1. Single story rural houses; usually modest and located in flat areas.
2. Two-storied semi-rural houses, common in urban centers of rural wards.
3. Two or three-storied urban buildings, located in towns.

Rural and semi-rural houses typically have external walls made of stone masonry and a wooden roof

structure. In some cases, internal walls are also made of stone masonry. Semi-rural houses have external walls supporting the floor, ceiling and wooden roof elements. Urban houses have stone masonry walls with large stones at the corners to improve the cross-link between the side walls. The internal walls are made of plastered lattices acting as building structural elements since they often contain a wood beam-column frame supporting several wood floors up to the roof.

The resistance of each house type depends on the disposition of the stones within the external walls. Costa and Arêde (2006) and Costa (2006) distinguish three different types of stone walls: first, high-quality walls, formed with a single stack of stones, which are often laid out in an angular manner, so that they wedge together, and are typically quite large. Second, irregular stone walls, built with "burnt stone" playing an important role to strengthen the structure. Voids are filled with small size materials or clay. Finally, there is the double-wall, made of selected stones slightly larger than half-thickness of the wall, where the stones are interlocked and form vertical layers. According to Costa (2006), the second and third types are predominant in rural areas, and these three wall types are highly vulnerable to ground shaking.

Reinforced concrete (RC) buildings found in the island are made of beams and columns connected by rigid joints, with or without infill walls, or with reinforced concrete walls. The typical infill wall material is concrete blocks. This RC type has one to six-stories, 1–2 stories being the most common for a single-family. Safety regulations and actions for building and bridge structures introduced in 1983 (Law no 235/83) have improved the RC frame building resistance to earthquake actions.

The different building types aforementioned are distributed into the vulnerability classes of the EMS-98 classification as follows:

- Reinforced concrete buildings in class D.
- Buildings with masonry walls of mortar with concrete slab built before 1971 are in class B, and in class C for those built between 1971 until 2001.
- Buildings with walls made of mortar without concrete slab are in class B.

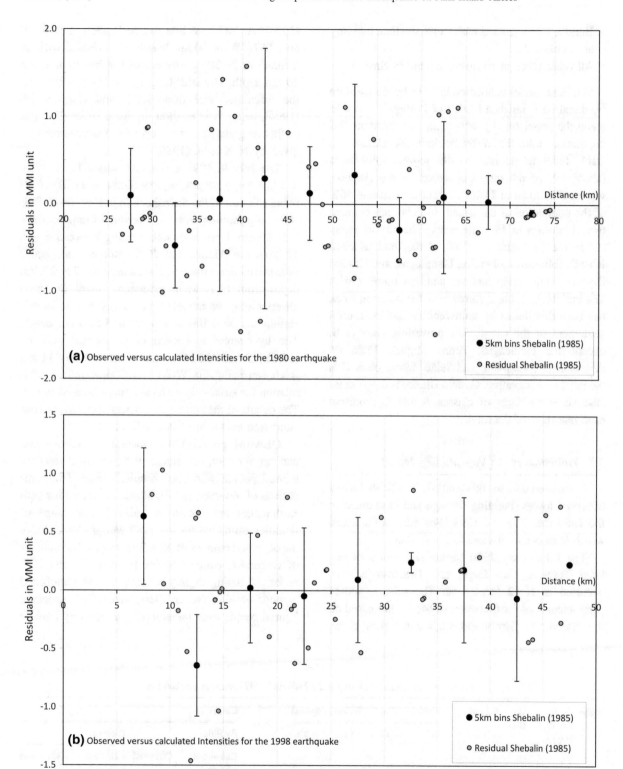

Figure 3
Observed versus calculated residuals for the **a** 1980 and **b** 1998 earthquakes. Black dots are the average values plus standard deviation grouped by 5 km epicentral distances derived from Shebalin (1985), and grey dots are the individual residual values

- Buildings constructed with adobe or stone masonry are in class A.
- All other types of constructions are in class A.

The last census conducted in 2011 by the Instituto Nacional de Estatística (2013) of Portugal is used to count the population by settlement and distribute this population into the different building classes. In 2011, Faial Island had 14,799 regular inhabitants (about 6% of the total population of the Azores), whereas the town of Horta concentrated around 40% of the population of the island. In Faial, it's counted 6447 buildings in 13 settlements. The city of Horta counted 2311 buildings (36% of the total of Faial Island), followed by Feteira, Flamengos, and Castelo Branco, while Salão had the smallest number with 174 buildings. Table 2 shows how the building stock has been distributed by vulnerability classes in each settlement of the Island. The prevailing class is B, except for Flamengos, Pedro Miguel, Praia do Almoxarife, Ribeirinha, and Salão, where class C is the most representative. Castelo Branco is a particular case since buildings of classes A and C represent each one-third of the total.

3.3. Validation of the Vulnerability Model

Calculated data for this analysis is validated using observed losses (building damage and casualties) in the 1980 (Ms 7.2) and 1998 (Mw 6.2) earthquakes, which both occurred close to Faial island.

The 1st January 1980 earthquake, which devastated Terceira, São Jorge, and Graciosa Islands, occurred at 16:42 local time. The seismic source parameters used for this event of Ms 7.2 are based on the aftershock distribution studied by Hirn et al.

(1980). An offshore line source located at 38.81°N and 27.78°W of 60 km length and oriented with an azimuth of N150 is selected, with a hypocenter at a 10 km depth. The fault length is an average between the calculated one from Wells and Coppersmith (1994), using their relationship for a strike-slip fault of this magnitude, and the estimate from aftershocks given by Hirn et al. (1980).

The July 9, 1998 earthquake shook Faial, Pico, and São Jorge Islands, at 5:19 (GMT − 1). The fault plane solution of the mainshock indicates NNW–SSE strike-slip faulting with the aftershocks ranging from 2 to 16 km depth and concentrating between 8 and 12 km depth (Matias et al. 2007). Borges et al. (2007) determined the rupture initiation between 7 and 8 km depth from body-wave inversion. Based on these observations, we modeled the energy release in this earthquake on a line source located at 5 km depth. The hypocenter is chosen to be located offshore (38.638°N and 28.524°W) and a line source of 11 km is chosen using the Wells and Coppersmith (1994) relation for strike-slip faults at a magnitude M of 6.2. The details of the source parameters selected for both validation events are listed in Table 3.

QLARM provides the calculated fatalities and patients for each settlement. Patients are defined as injured people who need hospitalization. The comparison of observed and calculated casualties for both earthquakes are shown in Table 3. The range of fatalities estimated for the 1980 earthquake includes the observed number of 61 killed people. The number of calculated patients varies between 38 and 285, under-estimating slightly the estimated and unofficial number of 300. Nevertheless, around 90 seriously injured people were mentioned in a report by USAID

Table 3

Parameters and scenario results for the 1980 and 1998 validation earthquakes

Name	Fault length (km) L	Z (km)	C2	M	Fault location (in decimal degrees)				Cal. Imax	Casualties			
					X1	Y1	X2	Y2		Fatalities		Patients	
										Calculated min–max	Observed	Calculated min–max	Observed
1980	60	10	5.2	7.2 Ms	− 27.878	38.967	− 27.682	38.653	VIII+	8–75	61	38–285	> 300
1998	11	5	5.7	6.2 Mw	− 28.544	38.675	− 28.504	38.601	VIII	3–15	8	7–111	128

(1980), which better fits the patients calculated in QLARM. The calculation done with population data from the 2001 census doesn't affect the results since the population growth of Terceira Island was around 4% between 1981 and 2001 (Table S1 of the electronic supplement of this article) which is a negligible variation compared to the uncertainties of the overall calculation.

For the 1998 earthquake, calculated fatality numbers, based on population data and building stock from the 2001 census, vary between 3 and 15 compared to the official death toll of 8 people. The calculated patients are in the range of 7–111, slightly underestimated compared to the official number of patients.

The two validation cases are satisfying both the correct estimate of the ground shaking field in terms of calculated intensity and the accurate estimation of the killed and seriously injured people on Faial

Island. It, therefore, justifies the use of QLARM to provide loss estimate for scenarios of likely future earthquakes.

3.4. Scenarios for Potential Earthquakes in Faial

The proposed earthquake scenarios are based on paleo-seismological studies by Madeira (1998) and Madeira et al. (2015) of Faial Island. These authors describe the seismogenic potential of active faults with magnitude, rupture length, and recurrence period. The two active faults selected to estimate losses in future likely earthquakes are the inland Espalamaca fault, hereafter called inland scenario, and the offshore fault between the Faial and Pico Islands called offshore scenario (as shown in Fig. 4). The input parameters used in QLARM to estimate losses are listed in Table 4. For both scenarios, the losses are calculated using a line source centered on

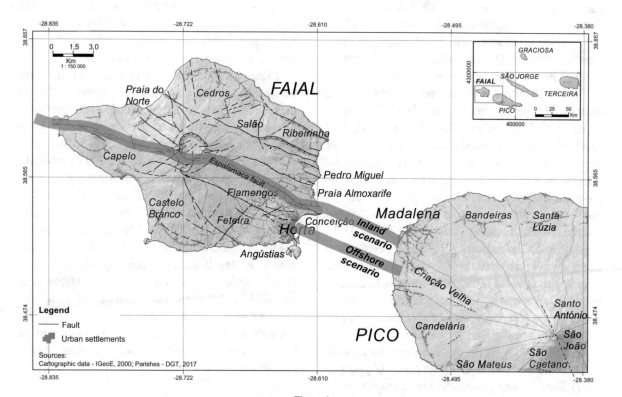

Figure 4

Location of faults selected (thick red lines) for earthquake scenarios. The number of fatalities and injuries requiring hospitalization is calculated for these two scenarios. The results are provided using the values 5.2 and 5.7 for the factor C2 in Eq. (1), given the uncertainty about the ground motions estimates observed in the 1980 and 1998 cases. Both of these estimates are based on worst case assumption of 90% of the population being inside during the night)

Table 4

Parameters and scenario results for the two scenarios proposed in Faial Island. Values are given for the nighttime case when 90% of the population is inside. C2 refers to the constant in Eq. (1). Z is the depth, L is the length of the fault. M is the magnitude. X1, Y1 and X2, Y2 are the coordinates of the fault ends. Imax is the maximum calculated EMS-98

Scenario	C2	Z (km)	L (km)	M	X1 (deg)	Y1 (deg)	X2 (deg)	Y2 (deg)	Imax calc	Fatalities			Patients		
										Mean	Min	Max	Mean	Min	Max
Offshore	5.2	5	8	6	− 28.625	38.537	− 28.536	38.509	IX	70	50	90	215	160	280
	5.7								X	155	125	190	445	355	545
Inland	5.2	5	27.5	6.9	− 28.842	38.600	− 28.540	38.530	XI	650	580	715	1770	1585	1945
	5.7								XII	635	570	700	1720	1530	1900

the epicenter with general NW–SE orientation, in agreement with the major tectonic structures of the island.

According to Madeira (1998) and Madeira et al. (2015), the extent of the Espalamaca fault is about 22.5 km. Three events have been identified during the last 4000 years generating displacements of 0.73 m, 1.65 m, and 0.46 m, respectively, with a return period estimated around 1300 years (Madeira (1998). We considered the maximum magnitude of 6.9 as given by these authors which correspond to a rupture length of 27.5 km using Wells and Coppersmith (1994) relationship for normal faulting.

The M6.9 inland Faial scenario strongly affects most of the settlements of the island, Flamengos being the village where the highest damage is expected because the calculated intensities are higher than XI due to site amplification. The population affected by intensity VI and higher is estimated to be on average 25,600, which is the total population of Faial and the western part of Pico Island. Average estimated fatalities and patients are 640 and 1750, respectively. The influence of the seismic wave attenuation is reduced in this case because the distances to affected settlements are low. Compared to nighttime scenarios, similar scenarios calculated for a daytime occupancy rate of 50% reducing by two the numbers of casualties, on average.

The offshore scenario is located on the seismogenic source of the 1926 earthquake (Agostinho 1927). For this zone, parameters such as rupture type, length, displacement and return period are unknown. A magnitude of 6 is assigned to the scenario which corresponds to a fault length of 8 km. The depth is

fixed at 5 km for each of the two scenarios since the seismogenic crust in the Azores varies between 8 and 11 km (Luis and Neves 2006), and agrees with the focal depths of the recent large earthquakes in the island (1926, 1958 and 1998). This scenario affects both Pico and Faial Islands with a maximum intensity of X in Flamengos. The population affected by intensities VI+ is estimated as about 22,500. Average numbers of estimated fatalities and patients are 110 and 330, respectively. Values of 95% and 5% for confidence levels are given in Table 4 for both applied C1 seismic attenuation factors.

Figure 5 shows, for each settlement of Faial, the percentage of building damage divided into six degrees from D0 (no damage) to D5 (very important with collapse) for the worst-case scenario of the M6.9 earthquake with low attenuation (C2 = 5.7). In general, the settlements affected by the 1998 earthquake will show strong resilience because many of those houses have been rebuilt. Flamengos is calculated to be the most damaged village because of its proximity to the epicenter. In other cases, houses located near or over-identified faults were built in new areas considered safe. The undamaged buildings in the city of Horta, Capelo and Praia do Norte, Pedro Miguel after the 1998 earthquake could be suspected to be less resistant during the next strong ground motion. On average, around 50% of the building stock will be heavily damaged or will collapse in an event of such magnitude.

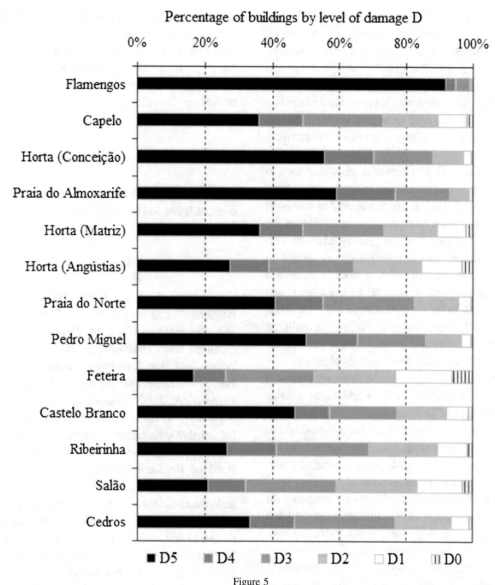

Figure 5

Building damage for the M6.9 scenario with C2 = 5.7. Settlements are ordered by increasing epicentral distance. Damage is divided into six degrees; D0: no damage, D1, slight; D2, moderate; D3, important; D4, very important without collapse; D5, very important with collapse

4. Discussion and Conclusion

A procedure has been applied to update and validate the QLARM software tool and database in the region of the Azores, Portugal. The global IPE from Shebalin (1985) implement in QLARM has been used to reproduce the intensity field for the 1980 and 1998 Azores earthquakes. The Shebalin relation has been used routinely for the alert service since 2003 (Wyss 2014) and generally well validated using past earthquakes in other regions of the world. In this particular case, the RMS error between observed and calculated intensities for distance up to 70 km is less than 0.7 which is much better than the result obtained with an attenuation relation from Munson and Thurber (1997) specifically developed for the volcanic Hawaiian Islands.

Soil conditions play a major role in the level of ground shaking in the Azores (Santos et al. 2011; Lopes et al. 2013; Veludo et al. 2013; Teves-Costa and Veludo 2013; Teves-Costa et al. 2008). The existing soil classification mapping (Forjaz et al. 2001) is limited to the Central group (Terceira, Graciosa, S. Jorge, Pico, and Faial Islands) and overestimates shear wave velocity for all site classes. To overcome this limitation, in settlements where site amplification is observed from past earthquakes, we added the same intensity increment to intensity calculations.

The 2001 and 2011 census data have been used to document settlement population change and has been extrapolated to 2016. Information related to houses in the census has been interpreted to characterize the Faial Island building stock regarding EMS-98 vulnerability classes for each settlement as shown in Table 2. On average, the building stock is composed of 8% of class A; 47% of class B; 30% of class C and 14% of class D. In the period between 2001 and 2011, the population decreased of 0.5% but the building stock in the same period increased 30% (INE 2013; the detail table is in the electronic supplement).

Two scenarios for likely earthquakes are proposed based on past earthquake activity and recent studies of active tectonics. The scenario of M6.9 is located inland on Faial, and the other scenario is assumed to be located offshore with M6. The average return period is approximately 1300 years for the first scenario and unknown for the second.

The inland scenario is the worst case proposed for Faial, yielding a range of fatalities between 580 and 710 and approximately three times more injuries. These numbers represents 3% and 5% of the total population of the island, respectively. The percentage of population affected by the earthquake due to building damage is almost the total population of the island; 35% of the buildings are estimated to be highly damaged or collapsed. For the offshore earthquake, the human losses are six times lower than for the onshore scenario, but the affected population by damage to houses remains high, around 24,000.

For the first time, QLARM is tested in the context of small volcanic islands with large heterogeneities in the geological processes, inducing specific soil condition and attenuation behavior. This work shows that QLARM gives satisfactory results to estimate human losses despite the limited knowledge of site effects. Human loss scenarios of future earthquakes are a valuable tool to encourage mitigation of the effects caused by such events and prepare emergency plans based on realistic scenarios. The seasonal variation of the population due to touristic activities is an issue that should be further investigated since the number of visitors is estimated at around 1.5 million each year.

5. Data and Resources

All data used in this paper came from published sources listed in the references.

Acknowledgements

The authors thank Dr. Goran Trendafiloski for his collaboration in the calibration phase of the project. We also thank António M. Medeiros of School of Sciences and Technologies of the University of Azores and Bob Bishop of International Centre for Earth Simulation Foundation for their comment and suggestions. João Fontiela is supported by grant M3.1.2/F/060/2011 of Regional Science Fund of the Regional Government Azores. This work is co-funded by the European Union through the European Fund of Regional Development, framed in COMPETE 2020 (Operational Competitiveness Program and Internationalization) through the ICT project (UID/GEO/04683/2013) with the reference POCI-01-0145-FEDER-007690 and by the bilateral project MITMOTION–Ground motion prediction in Mitidja Basin–Alger (PT-DZ/0003/2015).

Publisher's Note Springer Nature remains neutral with regard to jurisdictional claims in published maps and institutional affiliations.

REFERENCES

Agostinho, J. (1927). O terramoto de 31 de Agosto nas ilhas dos Açores. *Labor, 8,* 229–235.

Allen, T. I., Wald, D. J., Hotovec, A. J., Lin, K., Earle, P. S., & Marano, K. D. (2008). An Atlas of ShakeMaps for selected global earthquakes: U.S. Geological Survey Open-File Report 2008–1236 (p. 35).

Ambraseys, N. (1985). Intensity-attenuation and magnitude intensity relationships for northwest European earthquakes. *Earthquake Engineering and Structural Dynamics, 13,* 733–788.

Ambraseys, N. N., Simpson, K. A., & Bommer, J. J. (1996). Prediction of horizontal response spectra in Europe. *Earthquake Engineering and Structural Dynamics, 25,* 371–400.

Bird, P. (2003). An updated digital model of plate boundaries. *Geochemistry, Geophysics, Geosystems, 4*(3), 1027.

Boore, D. M., Joyner, W. B., & Fumal, T. E. (1997). Equations for estimating horizontal response spectra and peak acceleration from western North American earthquakes: A summary of recent work. *Seismological Research Letters, 68*(1), 128–153.

Borges, J. F., Bezzeghoud, M., Buforn, E., Pro, C., & Fitas, A. J. S. (2007). The 1980, 1997 and 1998 Azores earthquakes and some seismo-tectonic implications. *Tectonophysics, 435,* 37–54.

Buforn, E., Udias, A., & Colombas, M. A. (1988). Seismicity, source mechanisms and tectonics of the Azores-Gibraltar plate boundary. *Tectonophysics, 152*(1–2), 89–118.

Costa, A. (2006). Strengthening and repairing earthquake damaged structures. In C. S. Oliveira, A. Roca & X. Goula (Eds.), *Assessing and managing earthquake risk—Geo-scientific and engineering knowledge for earthquake risk mitigation: developments, tools, techniques. Geotechnical, geological and earthquake engineering* (Vol. 2, pp. 403–426). Dordrecht: Springer.

Costa, A., & Arêde, A. (2006). Strengthening of structures damage by the Azores earthquake of 1998. *Construction and Building Materials, 20,* 252–268.

Dias, N. A., Matias, L., Lourenço, N., Madeira, J., Carrilho, F., & Gaspar, J. L. (2007). Crustal seismic velocity structure near Faial and Pico Islands (AZORES from local earthquake tomography. *Tectonophysics, 445,* 301–317.

European Macroseismic Scale 1998 (EMS-98). (1998). EMS-98. In G. Grünthal (Ed.), Cahiers du Centre Européen de Géodynamique et de Séismologie (Vol. 15, p. 99).

Fernandes, R. M. S., Miranda, J. M. I., Catalão, J., Luis, J. F., Bastos, L., & Ambrosius, B. A. C. (2002). Coseismic displacements of the MW = 6.1, July 9, 1998, Faial earthquake (Azores, North Atlantic). *Geophysical Research Letters, 29*(16), 21.

Ferreira, M. A. (2008). Classificação dos danos no edificado com base na EMS-98. Sismo 1998–Açores. Uma década depois. Governo dos Açores/SPRHI S.A. (pp. 501–512).

Ferreira, M. A., & Oliveira, C. S. (2008). Overall impact of July, 9 Earthquake. Ten years later. In *Proceedings of AZORES 1998–international seminar on seismic risk and rehabilitation of stone masonry housing.*

Fontiela, J., Bezzeghoud, M., Rosset, P., & Cota Rodrigues, F. (2017). Maximum observed intensity map for the Azores Archipelago (Portugal) from 1522 to 2012 seismic catalog. *Seismological Research Letters, 88*(4), 1178–1184.

Forjaz, V. H., Nunes, J. C., Guedes, J. H., & Oliveira, C. S. (2001). Classificação geotécnica de solos das ilhas dos Açores: uma proposta. Actas do II Simpósio de Meteorologia e Geofísica (pp. 76–81).

Gonçalves, J. (2008). O atendimento hospitalar no sismo de 1998. In C. S. Oliveira, A. Costa & J. C. Nunes (Eds.), *Sismo 1998–Açores. Uma década depois. Governo dos Açores/SPRHI S.A.* (pp. 245–249).

Grimison, N. L., & Chen, W. (1986). The Azores-Gibraltar plate boundary: Focal mechanisms, depths of earthquakes, and their tectonic implications. *Journal of Geophysical Research: Solid Earth, 91*(B2), 2029–2047.

Grimison, N. L., & Chen, W.-P. (1988). Source mechanisms of four recent earthquakes along the Azores-Gibraltar plate boundary. *Geophysical Journal International, 92,* 391–401.

Hirn, A., Haessler, H., Trong, P. H., Wittlinger, G., & Mendes Victor, L. A. (1980). Aftershock sequence of the January 1st, 1980, earthquake and present-day tectonics in the Azores. *Geophysical Research Letters, 7*(7), 501–504.

Huo, J., & Hu, Y. (1992). Study on attenuation laws of ground motion parameters. *Earthquake Engineering and Engineering Vibration, 12,* 1–11.

Instituto Nacional de Estatística. (2013). Censos da População 2011. Lisboa: INE, i.P. Accessed Oct 2017.

Lagomarsino, S., & Giovinazzi, S. (2006). Macroseismic and mechanical models for the vulnerability assessment of current buildings. *Bulletin of Earthquake Engineering, 4*(4), 415–443.

Lima, G. (1934). Breviário Açoreano. Tip. Editora Andrade, Angra do Heroísmo (p. 399).

Lobão, C. (1999). O ano do vulcão: 1957–1958. Clube de Filatelia "O Ilhéu", Horta (p. 132).

Lopes, I., Deidda, G. P., Mendes, M., Strobbia, C., & Santos, J. (2013). Contribution of in situ geophysical methods for the definition of the São Sebastião crater model (Azores). *Journal of Applied Geophysics, 98,* 265–279.

Luis, J. F., & Neves, M. C. (2006). The isostatic compensation of the Azores Plateau: A 3D admittance and coherence analysis. *Journal of Volcanology and Geothermal Research, 156*(1–2), 10–22.

Machado, F. (1949). O terramoto de S. Jorge em 1757. *Açoreana, 4*(4), 311–324.

Machado, F., Parsons, W. H., Richards, A. F., & Mulford, J. W. (1962). Capelinhos eruption of Fayal Volcano, Azores, 1957–1958. *Journal of Geophysical Research, 67*(9), 3519–3529.

Madeira, J. (1998). Estudos de neotectónica nas ilhas do Faial, Pico e S. Jorge: Uma contribuição para o conhecimento geodinâmico da junção tripla dos Açores. PhD Thesis, Faculdade de Ciências da Universidade de Lisboa (p. 481).

Madeira, J., Brum da Silveira, A., Hipólito, A., & Carmo, R. (2015). Chapter 3: Active tectonics in the central and eastern Azores islands along the Eurasia–Nubia boundary: a review. *Geological Society, London, Memoirs, 44*(1), 15–32.

Matias, L., Dias, N. A., Morais, I., Vales, D., Carrilho, F., Madeira, J., et al. (2007). The 9th of July 1998 Faial Island (Azores, North Atlantic) seismic sequence. *Journal Seismology, 11,* 275–298.

Munson, C. G., & Thurber, C. H. (1997). Analysis of the attenuation of strong ground motion on the island of Hawaii. *Bulletin of the Seismological Society of America, 87*(4), 945–960.

Neves, F., Costa, A., Vicente, R., Oliveira, C. S., & Varum, H. (2012). Seismic vulnerability assessment and characterisation of the buildings on Faial Island, Azores. *Bulletin of Earthquake Engineering, 10*(1), 27–44.

Nunes, J. C., Forjaz, V. H., & França, Z. (2001). Principais sismos destrutivos no arquipélago dos Açores—Uma revisão. In M. R. Fragoso (Ed.), *5° Encontro Nacional de Sismologia e Engenharia Sísmica.* Ponta Delgada: LREC.

Nunes, J. C., Forjaz, V. H., & Oliveira, C. S. (2004). Catálogo Sísmico da Região dos Açores Versão 1.1 (1850–1998). In P. B. Lourenço, J. O. Barros & D. O. Oliveira (Eds.), *SÍSMICA 2004-6° Congresso Nacional de Sismologia e Engenharia Sísmica* (pp. 349–358).

Parvez, I. A., & Rosset, P. (2014). The role of microzonation in estimating earthquake risk. In J. Shroder & M. Wyss (Eds.), *Earthquake hazard, risk and disasters* (pp. 273–308). Boston: Elsevier.

Pena, J. A., Cruz, J., & Senos, M. L. (2001). Caracterização de efeitos de sítio nos Açores S. Sebastião e Flamengos. In Associação Portuguesa de Meteorologia e Geofísica (Ed.), *Actas do I Simpósio de Meteorologia e Geofísica—Comunicações de Geofísica* (pp. 88–94). Lagos (Algarve).

Rosset, P., Bonjour, C., & Wyss, M. (2015). QLARM, un outil d'aide à la gestion du risque sismique à échelle variable. In Plan de sauvegarde et outils de gestion de crise, F. Leone & F. Vinet (Eds.), *Presses Universitaires de la Méditerranée, Collection Géorisques* (Vol. 6, pp. 91–98).

Rosset, P., & Wyss, M. (2017). Seismic loss assessment in algeria using the tool QLARM. *Civil Engineering Research Journal, 2*(2), 555–583.

Santos, J., Chitas, P., Lopes, I., Oliveira, C. S., de Almeida, I. M., & Nunes, J. C. (2011). Assessment of site-effects using acceleration time series. Application to São Sebastião volcanic crater. *Soil Dynamics and Earthquake Engineering, 31*(4), 662–673.

Senos, M. L, Alves, P., Vales, D., Cruz, J., Silva, M., & Carrilho, F. (2008). O sismo de 9 de Julho de 1998 nos Açores e a crise sísmica associadas-dez anos depois. In C. S. Oliveira, A. Costa & J. C. Nunes (Eds.). *Sismo 1998–Açores. Uma década depois. Governo dos Açores/SPRHI S.A.* (pp. 73–87).

Senos, M. L., Gaspar, J. L., Cruz, J., Ferreira, T., Nunes, J. C., Pacheco, J., Alves, P., Queiroz, G., Dessai, P., Coutinho, R, Vales, D., & Carrilho, F. (1999). O Terramoto do Faial de 9 de Julho de 1998. In Associação Portuguesa de Meteorologia e Geofísica (Ed.), *Actas do I Simpósio de Meteorologia e Geofísica—Comunicações de Geofísica. Lagos (Algarve)* (pp. 61–67).

Shebalin, N. V. (1968). Methods of engineering seismic data application for seismic zoning. In S. V. Medvedev (Ed.), *Seismic zoning of the USSR* (pp. 95–111). Moscow: Science.

Shebalin N. V. (1985). Regularities of the natural disasters (in Russian), Nauki o zemle, Znanie (Vol. 11, p. 48).

Teves-Costa, T., Pacheco, J., Escuer, M., & COMICO Team. (2008). Utilização de vibrações ambientais na caracterização dinâmica das camadas superficiais na cidade da Horta. In C. S. Oliveira, A. Costa, J. C. Nunes (Eds.). *Sismo 1998–Açores. Uma década depois. Governo dos Açores/SPRHI S.A.* (pp. 137–149).

Teves-Costa, P., & Veludo, I. (2013). Soil characterization for seismic damage scenarios purposes: Application to Angra do Heroísmo (Azores). *Bulletin of Earthquake Engineering, 11*(2), 401–421.

Trendafiloski, G. S. (2007). *Earthquake casualty estimation, Summary research report*. Harbin: Institute of Engineering Mechanics (IEM).

Trendafiloski, G., Wyss, M., Rosset, P., & Marmureanu, G. (2009). Constructing city models to estimate losses due to earthquakes Worldwide: Application to Bucharest, Romania. *Earthquake Spectra, 25*(3), 665–685.

USAID. (1980). Disaster case report—The Azores earthquake (p. 6). http://pdf.usaid.gov/pdf_docs/PBAAH173.pdf.

Vacareanu, R., Lungu, D., Arion, C., & Aldea, A. (2004). Seismic risk scenarios, RISK-UE WP7 Handbook. RISK-UE Project: An advanced approach to earthquake risk scenarios with application to different European Cities. http://www.risk-ue.net.

Veludo, I., Teves-Costa, P., & Bard, P.-Y. (2013). Damage seismic scenarios for Angra do Heroísmo, Azores (Portugal). *Bulletin of Earthquake Engineering, 11*(2), 423–453.

Wald, D. J., Quitoriano, V., Heaton, T. H., Kanamori, H., Scrivner, C. W., & Worden, C. B. (1999). TriNet "ShakeMaps": Rapid generation of peak ground motion and intensity maps for earthquakes in southern California. *Earthquake Spectra, 15*(3), 537–555.

Wells, D. L., & Coppersmith, K. J. (1994). New empirical relationships among magnitude, rupture length, rupture width, rupture area, and surface displacement. *Bulletin of the Seismological Society of America, 84*(4), 974–1002.

Wyss, M. (2005). Human losses expected in Himalayan earthquakes. *Natural Hazards, 34*(3), 305–314.

Wyss, M. (2008). Estimated human losses in future earthquakes in Central Myanmar. *Seismological Research Letters, 79*(4), 520–525.

Wyss, M. (2010). Predicting the human losses implied by predictions of earthquakes: Southern Sumatra and Central Chile. *Pure and Applied Geophysics, 167*(8–9), 959–965.

Wyss, M. (2014). Ten years of real-time earthquake loss alerts. In J. Shroder & M. Wyss (Eds.), *Earthquake hazard, risk and disasters* (pp. 143–165). Boston: Elsevier.

Wyss, M. (2017a). Reported estimated quake death tolls to save lives. *Nature, 545*(7653), 151–153.

Wyss, M. (2017b). Four loss estimates for the Gorkha M7.8 earthquake, April 25, 2015, before and after it occurred. *Natural Hazards, 86*(S1), 141–150.

Wyss, M., Gupta, S., & Rosset, P. (2017). Casualty estimates in two up-dip complementary Himalayan earthquakes. *Seismological Research Letters, 88*, 1508–1515.

Wyss, M., Gupta, S., & Rosset, P. (2018). Casualty estimates in repeat Himalayan earthquakes in India. *Bulletin of the Seismological Society of America, 108*(5A), 2877–2893.

Wyss, M., Wang, R., Zschau, J., & Xia, Y. (2006). Earthquake loss estimates in near real-time. *Eos, Transactions American Geophysical Union, 87*(44), 477–478.

Wyss, M., & Wu, Z. (2014). How many lives were saved by the evacuation before the M7.3 Haicheng earthquake of 1975? *Seismological Research Letters, 85*(1), 226–229.

Wyss, M., & Zuñiga, F. R. (2016). Estimated casualties in a possible great earthquake along the pacific coast of Mexico. *Bulletin of the Seismological Society of America, 106*(4), 1867–1874.

Youngs, R. R., Chiou, S.-J., Silva, W. J., & Humphrey, J. R. (1997). Strong ground motion attenuation relationships for subducton zone earthquakes. *Seismological Research Letters, 68*(1), 58–73.

(Received May 23, 2019, revised September 10, 2019, accepted September 16, 2019, Published online September 23, 2019)

Printed in the United States
By Bookmasters